# SPICE

# A Guide to Circuit Simulation and Analysis Using PSpice®

# SPICE
# A Guide to Circuit Simulation and Analysis Using PSpice®

PAUL W. TUINENGA
*MicroSim Corporation*

PRENTICE HALL, *Englewood Cliffs, New Jersey 07632*

**Library of Congress Cataloging-in-Publication Data**

Tuinenga, Paul W.
SPICE: a guide to circuit simulation and analysis using PSpice/
Paul W. Tuinenga.
p. cm.
Bibliography: p.
ISBN 0-13-834607-0
1. SPICE (Computer program) I. Title.
TK454.T85 1988 88–2393
621.319′2—dc 19 CIP

Editorial/production supervision and
interior design: *Carolyn Fellows* and *Joseph Scordato*
Cover design: *Lundgren Graphics*
Manufacturing buyer: *Mary Noonan* and *Sallye Scott*



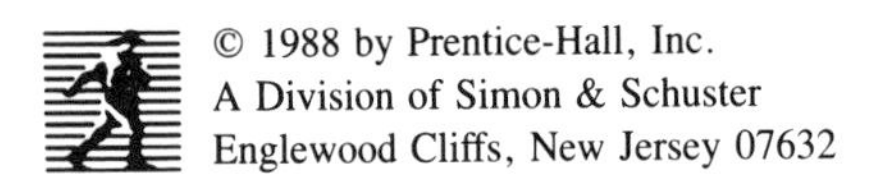



Printed in the United States of America
10 9 8 7

ISBN 0-13-834607-0

Prentice-Hall International (UK) Limited, *London*
Prentice-Hall of Australia Pty. Limited, *Sydney*
Prentice-Hall Canada Inc., *Toronto*
Prentice-Hall Hispanoamericana, S.A., *Mexico*
Prentice-Hall of India Private Limited, *New Delhi*
Prentice-Hall of Japan, Inc., *Tokyo*
Simon & Schuster Asia Pte. Ltd., *Singapore*
Editora Prentice-Hall do Brasil, Ltda., *Rio de Janeiro*

*To Linda, Claire, and William*

# Contents

# Preface

SPICE, from the University of California, at Berkeley, is the *de facto* world standard for analog circuit simulation. PSpice®, from MicroSim Corporation, is one of the many commercial derivatives of SPICE, and the first to be available on the IBM® personal computer. PSpice quickly became popular, and our customers found that there are no "how to" references for using SPICE, or its derivatives, the way there are references for database, spreadsheet, and word processing programs. This book's goal is to help users of PSpice, and many other SPICE-derived simulators, to access and use the features of the simulator for their work. Many of these features are only hinted at in other references, notes, or advice from other users, which are the traditional means of help you get for SPICE.

Beyond the syntax and semantics of the SPICE-standard input file, this book also demonstrates the use of PSpice for electrical engineering applications. There is a lot you can do with a circuit simulator that corresponds to what you might do on the lab-bench, as well as simulated measurements that go beyond what is possible with lab equipment.

The background required for using PSpice includes the following:

> This book assumes that you have a passing acquaintance with electrical or electronic circuits. This may come from formal study of basic electronic components (for instance, do you know what resistors and capacitors are, and how they react to electrical stimulation) and network analysis (for instance, do you recall Kirchhoff's laws). Or perhaps you picked up a working knowledge from your job, or as a hobby. If you have a circuit whose operation you basically understand, and you want to simulate this circuit to check the details of operation, you probably know enough to understand PSpice.

This book also assumes that you are able to operate the computer that will run the simulator, as well as create the input for the simulator. Perhaps a friend, or another kind person, will help you with this.

Many people made this book possible. Foremost, I thank my family for their patience while my preoccupation with this book kept me away from them. Also, I thank the people at MicroSim for their encouragement. Finally, I thank the users of PSpice whose many questions made the need for this book so obvious.

*Paul W. Tuinenga*

## ADDITIONAL ITEMS AVAILABLE

PSpice is available for the IBM-compatible personal computers, including the newer PS/2 series. PSpice is also available for workstations, minicomputers, and mainframes. Check with MicroSim for product availability. This book and the Student Version PSpice software are available in several formats:

1. A paperback edition with software included in the book (834614).
2. A paperback edition which does not include the software (834606).
3. The software is available separately in quantity from Prentice-Hall in an IBM® PC-compatible version (834630) a Macintosh II® compatible version (834622), and an IBM PS/2 compatible version (834648).

To order the software separately in quantity, please see the order card which is included in this book.

# Introduction

PSpice helps you simulate your electrical circuit designs **before** you build them. This lets you decide if changes are needed, without touching any hardware. PSpice also helps you check your design **after** you think it is complete. This lets you decide if the circuit will work correctly outside your office, in the real world, and have good production yield. In short, PSpice is a simulated ''lab bench'' on which you create test circuits and make measurements (PSpice will not design the circuit for you).

The practical way to check an electrical circuit is to build it. However, by the early 1970s, the components which were connected on an integrated circuit had become much smaller than individual discrete components. Physical effects that were negligible for normal circuits, such as a stereo amplifier, became important for these microcircuits. So the circuits could not be assembled from components in the lab and give the correct test results; the circuit had to be either (i) physically built, which is expensive and time consuming, or (ii) carefully simulated using a computer program. This is why the acronym, SPICE, stands for *Simulation Program with Integrated Circuit Emphasis.*

## WHAT IS PSPICE?

PSpice is a member of the SPICE ''family'' of circuit simulators, all of which derive from the SPICE2 circuit simulator developed at the University of California, Berkeley, during the mid-1970s. SPICE2 evolved from the original SPICE program, which, as it turns out, evolved from another simulator called CANCER that was developed in the early 1970s. Tremendous effort over this relatively short time created a simulator whose algorithms are robust, powerful, and general; SPICE2 quickly became an industry standard tool. Since this development was supported using public funds the software is "in the public domain," which means it may be freely used by U.S. citizens. The software is improved by U.C. Berkeley to the extent that it supports further research work. For example, SPICE3 is a "redesigned" implementation of the SPICE2 program that fits into U.C. Berkeley's computer-aided design (CAD) research program. SPICE3 is not better than SPICE2, in the way that SPICE2 was an advance over the original SPICE program; rather, it is designed to be a module in the U.C. Berkeley CAD research system. Neither SPICE2 nor SPICE3 is supported by U.C. Berkeley in the way commercial products are, nor does U.C. Berkeley provide consulting services. This led to commercial versions of SPICE which have the kind of support industrial customers require. Also, many companies have an in-house version of SPICE that has modifications to suit particular needs.

PSpice, which uses the same algorithms as SPICE2 (and conforms to its input syntax), shares this emphasis on microcircuit technology. However, the electrical concepts are general and are useful for all sizes of circuits (for example, power generation grids) and a wide range of applications. For instance, the simulator has no concept of large or small circuits; microvolts or megavolts are "just numbers" to PSpice. As long as PSpice is able to solve your circuit matrix, it will do so. This makes PSpice "technology independent" and generally useful. On the other hand, no assumptions are made about how the circuit should behave; for example, PSpice is not concerned that 0.03-watts output power does not make for a very loud stereo amplifier. You have to look at the results to see if they make sense in your application.

For discrete circuits (circuits made of individual parts assembled on a circuit board) PSpice has a variety of uses. Like the integrated circuit designs mentioned before, your designs are pressed for schedule time, budget expense, and manufacturing yield. With PSpice you can

Check a circuit idea before building a breadboard (even before ordering the parts).

Try out ideal, or "blue sky," operation by using ideal components to isolate limiting effects in your design.

Make simulated test measurements which are

- difficult (due to electrical noise or circuit loading),
- inconvenient (special test equipment is unavailable), or
- unwise (the test circuit would destroy itself).

Simulate a circuit many times with component variations to check what percentage will pass "final test," and find which combinations give the "worst case" results.

Once you become familiar with PSpice, you will find that it can substitute for most (but not all) of your breadboard work. Like any new tool, experience is required to get the most benefit from it.

The PSpice control statements (or "language," if you prefer) are easy to learn and use. These statements, which are collected in the file which is read by the simulator (the file is called a "circuit file"), are usually self-contained and may be understood without referring to any other statements. Moreover, each statement has so little interaction with other statements that they have the same meaning regardless of context. So the language of PSpice is easy to learn because you can focus on each statement type, master it, then move on to the next. Also, as you will see, you will not need to know many statement types to get started. Most of your difficulty will probably come from learning and operating your computer system.

## OTHER SPICE-BASED PROGRAMS

The commercially supported versions of SPICE2 fall into three groups:

The original group of mainframe-based versions, including HSPICE from Meta-Software, IG-SPICE from A.B. Associates, and I-SPICE from NCSS timesharing. HSPICE focuses on the needs of the integrated circuit designer with special device model support; Meta-Software also distributes RAD-SPICE which, as you might guess, simulates the operation of circuits subjected to ionizing radiation. I-SPICE and IG-SPICE focus on "interactive" circuit simulation and graphics output (which was an innovation for mainframe users). Precise from Electronic Engineering Software is a more recent addition to this group.

The IBM-PC based programs (besides PSpice), including AllSpice from Acotech, IS-SPICE from Intusoft, and Z-SPICE from Z-Tech. These are very similar to SPICE2; in most cases no changes have been made—even to correct errors (such as convergence problems). Without serious support for the simulator, these programs fall into a more "hobbyist" class of product. However, interesting additions include pre-processors or shell programs to manage input and provide "interactive" control, as well as post-processors for refining the normal SPICE output.

Advanced programs, with "innards" that are significantly overhauled, or entirely new, but adhering to the U.C. Berkeley standards for circuit description, including SPICE-Plus from Analog Design Tools, DSPICE from Daisy Systems, and PSpice from MicroSim. Many additions and improvements are available from these products. Also, these advanced simulators have options to extend simulation capabilities and interpret results.

The "growth" portion of the analog circuit simulation business is the last group, especially with the rapidly expanding market for engineering workstations. While U.C. Berkeley remains at the forefront of computer-aided tools for engineering, as a practical matter the complete simulation products will come from industry. Most, if not all, of the techniques you will see in this text are applicable to these products.

## ORGANIZATION OF THIS BOOK

This book adopts a "graduated example" (which some call a "tutorial") approach to learning about circuit simulation. It is tempting to load new software and try some examples, well before reading the instructions, so we will channel this urge toward learning about the simulator. We will start by building a simple circuit, make some DC "measurements," and move on. The biggest hurdle seems to be running a simulator the first few times. After that you start to focus on the electronics you are simulating and how best to measure what you want to discover.

The details of the semiconductor models are described later. These models are independent of the methods for using the simulator. In fact, we will do without them entirely in the examples.

An abridged summary of the control statements and device descriptions are at the end of the book. In fact, these appear as appendices because this information is not the *raison d'être* of the book, and is applicable to the SPICE-like simulators in only a general sense. You should try to obtain a detailed guide to the simulator you will be using.

Not "everything you ever wanted to know about SPICE . . ." is in this book. Missing are some topics that I had planned to include, as well as additional depth of coverage of the topics that do appear. These fell victim to the schedule for completing the text. Additional topics, without doubt, will become obvious from reader comments. Perhaps in the next edition. . . .

# SPICE

# A Guide to Circuit Simulation and Analysis Using PSpice®

# CHAPTER 1

# Getting Started

Let us begin with a quick circuit to introduce you to running PSpice. This will show you the basics of a circuit simulation without getting complicated by rules, details, exceptions, and so on, and quickly get to a successful result. Later, we will get to the wide range of features and ways of combining these to express complex circuit functions.

Sometimes the examples will omit features of the simulator intentionally to concentrate on a particular topic. These features are necessary for normal use, and we will get to them in due course. The examples are brief, to demonstrate an idea and there is the danger that, in not explaining somewhat unrelated items, they may mislead you. If this happens it was not intended, but is just a problem with this approach.

Sometimes the examples will repeat some of what was done already.

## 1.1 A SMALL CIRCUIT

The best way to learn a circuit simulator is to ''do'' simulations. Usually you start with a small circuit that you know, by inspection, will work.

Running this simulation requires several basic accomplishments. It requires that you (i) create the input file, or ''circuit file'' (although some call it a ''program'' for the simulator), (ii) run the simulator (without errors), (iii) find where the output went, and (iv) inspect the output.

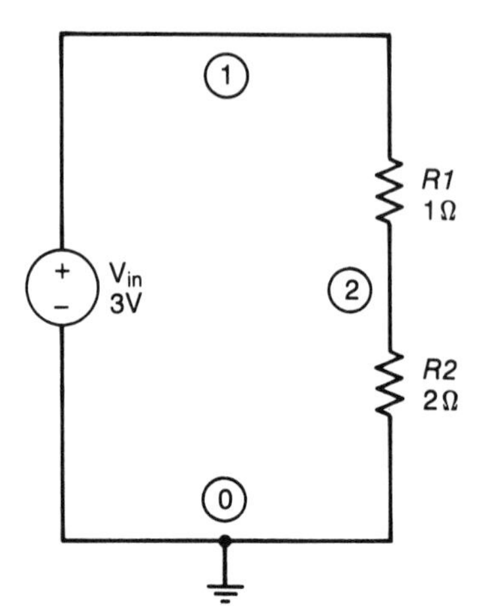


**Figure 1.1** Schematic for small-circuit example.

In PSpice, the circuit file to simulate this circuit is

```
* Resistor divider circuit
VIN 1 0 3.0volt
R1  1 2 1.0ohm
R2  2 0 2.0ohm
.END
```

How you run this simulation will depend on the system you are using. You will need to learn to use a text editor to create the input file. Then you will run PSpice, specifying the input file you created. If everything works, PSpice will read your input file called, for example, TEST.CIR, and place the results in an output file called TEST.OUT. The same text editor you used for creating the input file can also be used to inspect the output file. This output file may also be directed, by you, to a printer.

**Exercise 1.1.1**

Create and run this simulation on your system. Look at the output file. Did the printout show (correctly) node 2's voltage as 2 volts? Experiment by changing the circuit file—leave out, or add, something and see what errors PSpice will check.

Now to describe, and explain the circuit file. PSpice always expects the first line of the circuit file to be a title line. You can leave it blank, but circuit description can not start until the second line of the file. The examples in this book will sometimes start the title line with a "*" (which also indicates a comment line) by force of habit on my part. This is not necessary. What is necessary is the last line, ".END", which completes the description of the entire circuit including any simulation controls. You use ".END" because PSpice will let you start another, completely different, circuit simulation right after ".END". Between the first and last line, the circuit file may be in almost any order.

All of the circuit elements, or devices, in the circuit file are connected (in the sense that you would solder their leads together) by circuit nodes. You may think of these nodes as the connecting wires, or lines, in a circuit schematic. In SPICE2 these nodes are positive integers, including 0 (zero) which is reserved to mean "ground." PSpice does not require that you use integers (any text string will do), but 0 is still "ground." Every circuit file must have a ground node, as a reference, and every other node in the circuit file must have a DC current path to ground. This is one of the requirements of the SPICE algorithms.

Along with requiring a ground node, PSpice also requires that all terminals be connected to at least one other terminal. This is a precaution against dangling wires. Even though you may do this on the lab-bench, it is considered an error by the simulator.

The circuit file for our example uses only two-terminal devices—a voltage source and the resistors. A separate line is used to describe each element in the circuit. The basic syntax is

*<name> <node> <node>* . . . *<value>*

There are no one-terminal devices in PSpice. Devices with more than two terminals use basically the same form, but with more node items. The device "value" is a number, either decimal or floating point, that describes the size of the device. You will see later that there are a variety of ways to express the same value, including a metric suffix. After the value you may include a unit, such as "volt" or "ohm," for your own use; PSpice actually ignores these (to the extent that they aren't confused with one of the metric suffixes).

Any line may be a comment line by starting it with a "*" in the first column. This allows you to document your circuit file for others, unfamiliar with the circuit file, or for yourself when, after some time, you too will be unfamiliar with the circuit file. Blank lines are ignored; use them to separate sections of your circuit file.

PSpice also allows you to insert comments on any line by starting the comment with a ";" (semicolon). Everything on the line after the ";" is ignored; for example:

```
Rbias 2 3 45 ; this is the biasing device and had better not fail!
```

**Exercise 1.1.2**

Modify the previous exercise's circuit file by swapping any of the lines (except the first or last line). See if PSpice gives different results.

**Exercise 1.1.3**

Try adding some comment lines and blank lines to the previous exercise's circuit file. Then, try changing all of the 0 nodes to 3 (so there is no "ground") and see what happens. Then, try disconnecting R1 and R2 by adding a new node to the circuit file and see what errors result.

Take a look at the output file of our example. As it turns out, this example did not specify any type of simulation, such as frequency response, however PSpice assumes that at least you wanted a DC bias-point to be calculated. This is a calculation of what voltages the nodes would have if the circuit is quiescent, which also means the currents through the devices are calculated. These are printed in your output file. In addition, PSpice checks all of the devices which supply current to the circuit and totals the quiescent power dissipated by the circuit.

## 1.2 COMPONENT VALUES

All of the quantities, or values, in PSpice may be expressed as decimal or floating point values traditionally used by computer programs. The decimal numbers should be familiar; for example:

1 3.14 −13.7 .0045

Floating point values scale a decimal number by a power of ten, where the letter "E" (for "exponent") separates the decimal number from the start of the integer exponent, so that

.0045 can be written 4.5E-3 which means $4.5*10^{-3}$

Older SPICE versions also allow you to use "D" instead of "E." This is a holdover from the FORTRAN programming language where "D" meant that the number was stored with greater precision ("double" precision). PSpice will also accept the "D" format, but the storage precision is selected depending on the needs of the simulator.

Also, PSpice lets you use a metric-like suffix to express a value. These suffixes multiply the number they follow by a power of ten (with one exception). Using the suffix notation allows values written into the circuit file to look like the values on a circuit schematic. This is a great convenience that removes a source for most simulation errors: using the wrong component values.

These are the power-of-ten suffix letters, along with the metric prefixes and scale factors they represent, used by PSpice:

| | | |
|---|---|---|
| F | femto | $10^{-15}$ |
| P | pico | $10^{-12}$ |
| N | nano | $10^{-9}$ |
| U | micro | $10^{-6}$ |
| M | milli | $10^{-3}$ |
| K | kilo | $10^{+3}$ |
| MEG | mega | $10^{+6}$ |
| G | giga | $10^{+9}$ |
| T | tera | $10^{+12}$ |

plus this suffix (for English-to-metric conversion of integrated circuit device sizes):

MIL $25.4 \cdot 10^{-6}$

Using the exponential and suffix notation lets you express the same value many ways; for example:

1050000    1.05E6    1.05MEG    1.05E3K    .00105G

are all the same value to PSpice.

The previous list of suffixes was written in capital letters because the original SPICE simulators allowed only capital letters in the circuit file. One problem with this is the confusion between the standard use of "m" for "milli-" and "M" for "mega-," which was resolved by requiring the input to be "MEG" for "mega-." To maintain compatibility PSpice still requires that you use "MEG" (or "meg") for "mega-."

Other letters (those that are not suffixes) may be used with a number, but these are ignored by PSpice. That is why you may write "10," or "10volts," or "10ohms," to make your circuit file more readable without changing the meaning of the value. Moreover, once a valid suffix is read by PSpice, the remaining letters are ignored. You may also write "10pF," or "10picoamps," or "10picoseconds," and these will all be the value "10E-9."

# CHAPTER 2

# DC Operation

In the previous chapter PSpice calculated the DC bias-point for the circuits you entered. PSpice must do this before proceeding to any other type of analysis, since it must determine the operating point of the circuit (the voltages at each node and currents through each device). If you were to physically build a circuit and attach a power supply to it, when you start the supply the circuit will bias itself at its DC operating point. For most circuits this is a stable condition, without oscillation, and PSpice will arrive at a DC solution to the circuit. Later we will cover more stubborn circuits. These circuits work on the lab-bench, but PSpice will need help to calculate the DC bias-point.

While PSpice always calculates a bias-point before proceeding, it will not print out the results of this calculation unless (i) there are no other types of analyses specified, or (ii) you include a .OP statement in your circuit file. Even if you want only the DC bias information it is helpful to include the .OP statement to remind you later that the DC bias information was what you were after in the circuit. In a sense, .OP is one of the analyses that PSpice will perform on your circuits.

Using just the DC bias-point analysis, we can demonstrate some electrical laws that PSpice follows in calculating voltages and currents. But first, let us learn more about the basic components for building circuits.

## 2.1 PASSIVE DEVICES

The passive devices are resistors, capacitors, and inductors:

Resistors limit (resist) the flow of electrical current, following the law $V = I \cdot R$, where $V$ is the voltage (in *volts*) across the resistor, $I$ is the current (in *amperes*) through the resistor, and $R$ is the resistance value (in *ohms*).

Capacitors store energy in an electrostatic field, following the law $Q = V \cdot C$, where $Q$ is the induced charge (in *coulombs*) on the "plates" of the capacitor, $V$ is the voltage (in *volts*) impressed on the "plates," and $C$ is the capacitance value (in *farads*).

Inductors store energy in an electromagnetic field, following the law $\lambda = I \cdot L$, where $\lambda$ is the induced magnetic flux (in *Weber-turns*) around the inductor, $I$ is the current (in *amperes*) through the inductor, and $L$ is the self-inductance value (in *henries*).

Fortunately, most of the Rs, Ls, and Cs we use on the lab-bench are nearly ideal and for our purposes we can consider them to be ideal. In PSpice we can specify these devices merely by using the appropriate letter as the first letter of the device name:

Rxx for resistor
Lxx for inductor
Cxx for capacitor

and the xx represents any other letters or numbers you want to use to finish the "name" of the device.

To specify the device in the circuit file we include the name of the device, how it is connected into the circuit, and its value. PSpice uses the basic electrical units for voltage (volts) and current (amps) and also uses the basic electrical units for device values: ohms, farads, and henries.

Here are some example devices:

| | | |
|---|---|---|
| R12 | 5 2 15K | is a 15-kilohm resistor (15,000 ohm) |
| C2 | 12 3 1.8u | is a 1.8-microfarad capacitor (0.0000018 farad) |
| L3 | 7 6 10m | is a 10-millihenry inductor (0.01 henry) |

## 2.2 COMPONENT NAMES

As you just saw above, the names for devices started with an alphabetic letter reserved for that device. It is the first letter that tells PSpice what kind of device you are about to describe. These letters correspond to the standard ones used on circuit schematic diagrams for labeling devices. For instance, if you used "R17" as the label of a resistor in a schematic then you would probably use "R17" in your circuit file as the name of that resistor. The remaining letters of the component name may be alphabetic, or numbers, or (in PSpice) an underscore "_" or dollar sign "$" character. Upper- or lower-case letters may be used, but PSpice is not sensitive to which case is used, so that

RBIAS
Rbias
rbiaS

all refer to the same device. The maximum length possible for component names is longer than 80 characters. Practically, the length of the name you use for a component depends on how much typing you want to do.

Older SPICE versions, due to limitations of the computer language they were written in, as well as the need to conserve memory on the (then) current generation of machines (when "core" memory referred to the tiny, hand-strung, magnetic rings which stored only one bit), limited component names to eight characters. Of course, the first character specified the component type, so there were only seven characters left for making the name unique and identifiable.

Some of the statements in SPICE can be detailed and long, especially in PSpice where you are allowed to have long names for devices and nodes. As a convenience you may split a line, wherever you could normally use a space character, and continue on the next line. However, the first character on the continuation line must be a "+" to indicate that it is a continuation line; for example:

```
ResistorWithLongName ConnectedToOneNode AndToAnotherNode
+120ohms
```

## 2.3 INDEPENDENT SOURCES

To simulate your circuits you will also need some way to tell PSpice what is "exciting" or supplying electrical power to the circuit. For this we use independent sources which supply a fixed voltage level or current flow. We specify these sources in a way that is similar to the passive devices described earlier: name, connecting nodes, value. As you might have guessed

Vxx is a voltage source, and

Ixx is a current source

Remember, PSpice uses the basic electrical units for values; so the following examples are easy to understand:

| | |
|---|---|
| VIN 3 0 1.2K | is a 1.2-kilovolt source (1,200 volts) |
| I4 12 2 15m | is a 15-milliamp source (0.015 amps) |

A voltage source is like a battery, or lab-bench power supply. Current flows (using the positive current convention) out from the positive terminal (first node), through the circuit, and then into the negative terminal (second node). This is the conventional current flow taught to students. But why, in our first example, did PSpice calculate the supply current as a negative value? Because whenever you ask PSpice to print the value of a current through a device, "through" means into the first terminal and out the second terminal. In this case the current was flowing out of the positive (first) terminal, so the current has a negative value.

A current source provides a fixed value of current to the circuit. However, its current flows (again using positive current) into the positive terminal (first node), through the source, and then out of the negative terminal (second node). This is the opposite direction of the voltage source, but is consistent with reporting the current through the device. Also, current flows from the more positive potential to the more negative potential—in this case through the current source device.

**Exercise 2.3.1**

Using the circuit from the first exercise, replace the voltage source VIN with a current source of value 1 amp. How did the output from PSpice change?

## 2.4 OHM'S LAW

PSpice calculates values according to many laws of physics including Ohm's law, which we have seen already. The output from your first exercise showed the current flowing through the resistors and the voltages across each. Take a moment to check that Ohm's law was followed for each resistor.

## 2.5 KIRCHHOFF'S NETWORK LAWS

Consider the circuit in Figure 2.1,

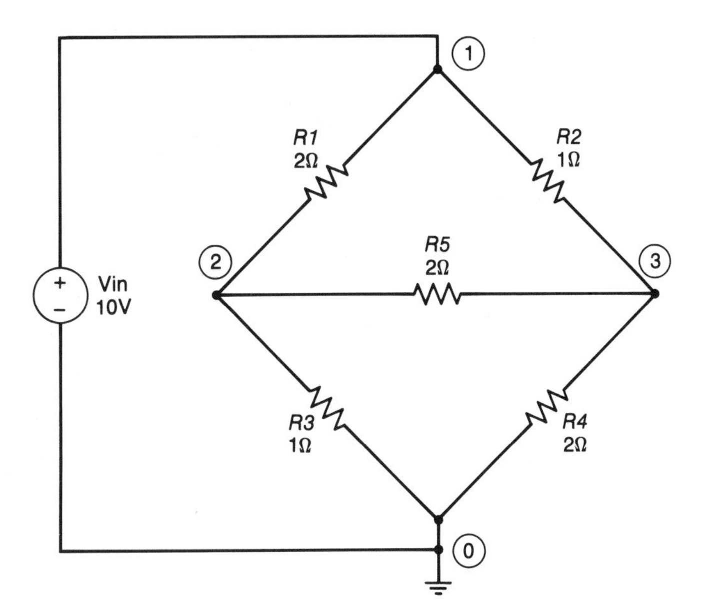


**Figure 2.1** Schematic for resistor bridge circuit.

with the equivalent circuit file:

```
Resistor bridge
VIN 1 0 10
R1  1 2 2
R2  1 3 1
R3  2 0 1
R4  3 0 2
R5  3 2 2
.End
```

After you run PSpice on this circuit, add up the voltage drops around any of the loops in the resistor network. For example, the loop of resistors R2, R5, and R1, have voltage drops (going clockwise) of 4 volts, 2 volts, and −6 volts. The sum of these voltages is always zero, which demonstrates one of Kirchhoff's network laws: *the algebraic sum of the potential drops around any closed loop in a network of conductors is always zero.*

**Exercise 2.5.1**

Sum the voltage drops around these loops: R3+R5+R4 and R1+R2+R3+R4.

Now we are going to try something a little different. Earlier we saw that PSpice outputs the current through the voltage source. If we use a voltage source with a value of 0 volts, we can insert this into the circuit and measure (like an ammeter) currents flowing through the circuit. Let us change the circuit to measure some currents, as in Figure 2.2,

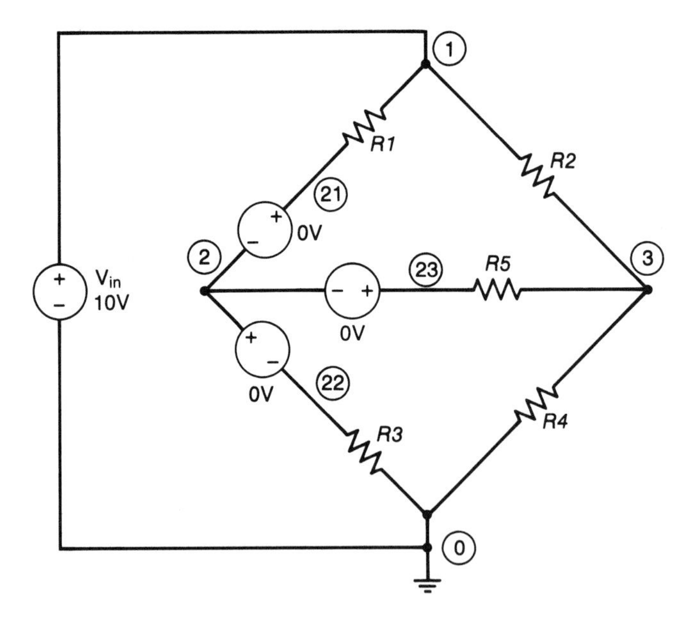


**Figure 2.2** Schematic for resistor bridge circuit, including zero-volt sources.

with the equivalent circuit file:

```
Resistor bridge
VIN 1 0 10
R1  1 21 2
V1  21 2 0
R2  1 3 1
V3  2 22 0
R3  22 0 1
R4  3 0 2
R5  3 23 2
V5  23 2 0
.END
```

After you run PSpice on this circuit, add up the currents for V1, V3, and V5. These turn out to be zero, which demonstrates one of Kirchhoff's network laws: the algebraic sum of the currents coming into any junction in the network is always zero.

**Exercise 2.5.2**

Change the polarity of one of the zero-volt sources and see how the output from PSpice changes. Does Kirchhoff's law still hold? Now, try inserting zero-volt sources around node 3 and check the output from PSpice.

## 2.6 CAPACITORS IN DC CIRCUITS

Capacitors block sustained, or DC, current. The only time current flows through the capacitor is when charge is collecting on, or being removed from, the "plates." This means that the voltage across the capacitor is changing, which is not the DC case. Of course in a real circuit, once the power is supplied, there is a transient during which some capacitors will charge up to their final values. But the result is the same as if these capacitors did not exist and the connections to each capacitor were left dangling. Inside PSpice, this is exactly how capacitors are treated for DC calculations.

Since the capacitors in PSpice are perfect (that is, without any leakage) it is important that there are no sections of your circuit that become isolated by "ignoring" the capacitors. This means that every node of your circuit needs to have some path for DC current, however convoluted, to ground so that bias levels may be determined for every node. If you have a circuit with a node that is isolated by perfect capacitors, for example, node 1 in the following circuit fragment:

```
C1 1 0 1pF
C2 1 2 1pF
V1 2 0 3volts
```

then there is no way, in theory, to determine the DC level of this node. In a real circuit, leakages in the dielectric of the capacitor would prevent such node from attaining a zillion volts.

PSpice checks for isolated nodes before starting any simulations. If you accidentally isolate a node you will receive an error message similar to

```
No DC path to node . . .
```

to indicate this problem. This simulation cannot proceed until this is corrected.

For the example just shown, you would want to connect a large-value resistor, say 1-gigaohm, to the isolated node. The other end of the resistor would be connected to ground, or whatever voltage level you wanted to use as the bias level.

**Exercise 2.6.1**

Insert a capacitor into one of the "legs" of the resistor bridge circuit (see earlier exercises). Run PSpice and note how the bias level changes. Try different "legs" of the bridge and check that the results were as you expected.

**Exercise 2.6.2**

By inserting two capacitors, isolate a resistor in one of the "legs" of the resistor bridge circuit (see earlier exercises). Run PSpice and find the error message caused by this situation.

## 2.7 INDUCTORS IN DC CIRCUITS

Inductors, which are essentially coils of wire, conduct DC current so they do not have the same restrictions as capacitors (previously discussed). However, they have a different restriction (of course, there's always a catch!) because of how inductors are simulated when the analyses includes time, such as AC response or transient simulation. In these cases the inductor develops a voltage across its winding in response to the changing magnetic flux within the windings. The total voltage developed is the sum of (i) flux changes due to current in the winding itself, and (ii) flux changes due to current in any other winding whose magnetic field is coupled to the winding in question. The reciprocal ratio of the change in current to the developed voltage is called, for these two cases, "self inductance" and "mutual inductance," respectively.

These voltages which are developed by flux changes are modeled, in PSpice, as time varying voltage sources. So far this is not a problem, except when you try to connect an inductor directly across a voltage source, even if the voltage source's value is zero. This situation is called a "voltage loop" which, as it sounds, means a circular path of voltage sources without any intervening resistance to limit the current to a finite value. To strictly check for all voltage loops, PSpice treats inductors as though they were voltage sources. For the DC case they are "shorts" or zero-volt sources, and for a time-related case they may be non-zero sources at some instant.

Furthermore, you may not connect two inductors in parallel. For the same, strict reason, each inductor is considered to be a voltage source so the parallel connection of two, or more, forms a "voltage loop." You may circumvent this restriction by including a series resistor with each inductor to "break" the loop. This may be a resistor of negligible value, say 0.001 ohm, or one which accounts for the winding resistance (the DC resistance of the coiled wire) in which case it will have the same resistance value as the winding.

**Exercise 2.7.1**

Insert an inductor into one of the "legs" of the resistor bridge circuit (see earlier exercises). Run PSpice and note if the bias levels change. Try different "legs" of the bridge and check that the results were as you expected.

**Exercise 2.7.2**

By inserting two inductors in parallel, create a voltage loop in one of the "legs" of the resistor bridge circuit (see earlier exercises). Run PSpice and find the error message caused by this situation.

# CHAPTER 3

# DC Sensitivity

Simulators are generally used either to verify a design, or to refine (improve) a design. Verifying is simply checking that the design "meets spec." However, refining the design may make it more robust, attain a "tighter spec," or even make it less expensive to produce. DC sensitivity calculations help guide the user to those components which affect a circuit's DC bias-point the most. This then will focus efforts on reducing the sensitivity of the circuit to component variations and/or drift, or it may provide evidence that a design is too conservative and that less expensive components, with more variation and/or drift, may be used.

## 3.1 THE .SENS STATEMENT

The ".SENS" statement specifies which DC outputs you want to consider (PSpice doesn't know how your circuit is being used). Then, once the DC bias-point for the circuit is calculated, PSpice calculates the sensitivity of each output, individually, to all of the device values (as well as "model parameters," which we will cover in due course) in the circuit. The format for the statement is

.SENS *<output value>* . . .

The *<output value>* is in the same format as for the .PRINT statement (see section 4.3, page 20).

## 3.2 DC SENSITIVITY ANALYSIS

Having a .SENS statement in your circuit file causes the DC sensitivity calculations to be done when the DC bias-point calculations are completed. You do not need to specify any other output to get the results of the DC sensitivity analysis. We can now try working some examples to see what .SENS will do. Usually you will be using .SENS to analyze a more complicated, active device circuit, such as a transistor amplifier. However, for ease of understanding we can demonstrate the use of .SENS with a small demonstration circuit, and then work an example showing a practical application.

Consider the simple circuit,

```
Resistor divider
VIN  1 0 1volt
R1   1 2 3ohm
R2   2 0 1ohm
.SENS V(2)
.END
```

which we will analyze. PSpice is run, and in the output file will be something similar to

```
DC SENSITIVITIES OF OUTPUT V(2)

                                 ELEMENT          NORMALIZED
ELEMENT       ELEMENT          SENSITIVITY        SENSITIVITY
 NAME          VALUE          (VOLTS/UNIT)     (VOLTS/PERCENT)

  R1         3.000E+00         -6.250E-02        -1.875E-03
  R2         1.000E+00          1.875E-01         1.875E-03
  Vin        1.000E+00          2.500E-01         2.500E-03
```

There will be a table like this for each <*output value*> in the .SENS statement. Also, the sensitivities to selected currents would be labeled "AMPS/UNIT" and "AMPS/PERCENT."

What does this mean? First let's look at the results for VIN. Our resistor divider has a voltage "gain" of ¼, that is, the variation of the voltage at V(2) is one-quarter of the variation in VIN. This results in a sensitivity of 0.25 volt change in V(2) for a one-volt change in VIN, or 0.25 volts/unit (the "unit" for a voltage source being volts). How can we check this calculation? The .SENS statement also caused the output of the bias-point calculation. We can see that the ratio of V(2) to VIN is ¼; the circuit has only linear elements, so doubling the value of VIN will double the value of V(2).

The last column shows sensitivity normalized to the component value, that is, as a percentage change. These values are then calculated by multiplying the former column (volts/unit) value by the element value, and then dividing by 100 to obtain a percent value.

By looking at the values calculated for R2, which (again) are equal since the resistor's value is one ohm, we can verify these by considering that

$$V(2) = VIN \cdot R2/(R1 + R2)$$

This means that the sensitivity of V(2) to R2 will be

$$(d/d\text{R2}) \cdot \text{VIN} \cdot \text{R2}/(\text{R1} + \text{R2})$$

which works out as

$$\text{VIN} \cdot ((\text{R1} + \text{R2}) - \text{R2})/(\text{R1} + \text{R2})^2 = \text{R1}/(\text{R1} + \text{R2})^2 = 3/16$$

By analogy we can calculate the sensitivity of V(2) to R1 will be

$$\text{VIN} \cdot ((\text{R1} + \text{R2}) - \text{R1})/(\text{R1} + \text{R2})^2 = \text{R2}/(\text{R1} + \text{R2})^2 = 1/16$$

Notice that for R1 the normalized sensitivity is identical, in magnitude, to the normalized sensitivity for R2. It follows that a percentage change in either section of the resistor divider would produce the same size of effect on the output. However, increasing R1 will decrease the output voltage so the sensitivity values are negative for this resistor.

**Exercise 3.2.1**

Run the sensitivity analysis just described. Now, change the value of R1 to 4 ohms and the value of VIN to 2 volts. Are the results what you expected?

## 3.3 CIRCUIT EXAMPLE: WORST-CASE DESIGN

One type of circuit where sensitivity to element values is of great importance is the digital-to-analog converter. These circuits generally use component ratios to

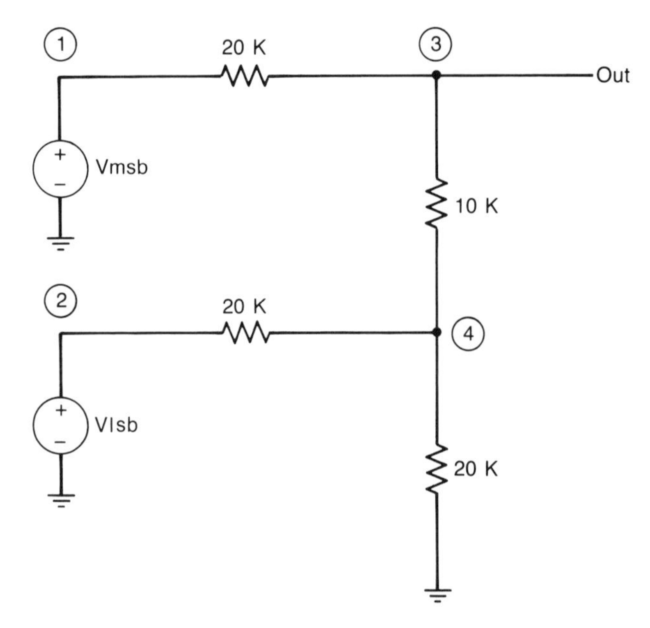


**Figure 3.1** Schematic for D-to-A converter using an R-2R ladder network.

generate a voltage that is a fraction of a reference voltage, where the fractional amount is set by a digital (binary coded) input. Digital systems, such as computers, may generate analog signals by using these circuits. Consider this circuit, represented by the following circuit file:

```
* Sensitivity analysis of D=A converter
Vmsb 1 0 0volt; most-significant-bit input
Vlsb 2 0 0volt; least-significant-bit input
R1 1 3 20K
R2 2 4 20K
R3 3 4 10K
R4 4 0 20K
.sens v(3)
.end
```

The inputs to this circuit are set at either 1 volt, or zero, depending on the binary input we are simulating. In this type of converter, called an R-2R ladder because of the resistor ratios, the input voltages are both the binary input as well as the reference voltage. You may think of it as being the binary bit value, 1 or 0, multiplied by the reference voltage. The output voltage, at node 3, is a fraction of the reference voltage, controlled by the ratio of the current binary input value to the number of values representable. In this case, with 2 bits, we will be able to generate the following fractions: $0/4$, $1/4$, $2/4$, and $3/4$.

To generate all of the input (binary code) cases we will need to make four runs. Each run will use a different combination of Vmsb and Vlsb, with values of either 0 volt or 1 volt, in the following combinations: 0-and-0, 0-and-1, 1-and-0, and finally 1-and-1. This provides the binary input for the decimal numbers 0, 1, 2, and 3. The output at V(3) will DC-bias to the voltage levels of 0, 0.25, 0.5, and 0.75, for these inputs, respectively. Looking at the sensitivity table for the inputs 0-and-0:

```
DC SENSITIVITIES OF OUTPUT (V3)

                                 ELEMENT          NORMALIZED
ELEMENT        ELEMENT         SENSITIVITY        SENSITIVITY
 NAME           VALUE          (VOLTS/UNIT)     (VOLTS/PERCENT)

 R1           2.000E+04         0.000E+00          0.000E+00
 R2           2.000E+04         0.000E+00          0.000E+00
 R3           1.000E+04         0.000E+00          0.000E+00
 R4           2.000E+04         0.000E+00          0.000E+00
 Vmsb         0.000E+00         5.000E-01          0.000E+00
 Vlsb         0.000E+00         2.500E-01          0.000E+00
```

The results are rather boring. Since all of the node voltages are zero it is difficult to affect the output voltage by changing a resistor value. Moving on, we look at the results for the combination 0-and-1:

| | | | |
|---|---|---|---|
| R1 | 2.000E+04 | 6.250E-06 | 1.250E-03 |
| R2 | 2.000E+04 | -7.813E-06 | -1.563E-03 |
| R3 | 1.000E+04 | -6.250E-06 | -6.250E-04 |
| R4 | 2.000E+04 | 4.688E-06 | 9.375E-04 |
| Vmsb | 0.000E+00 | 5.000E-01 | 0.000E+00 |
| Vlsb | 1.000E+00 | 2.500E-01 | 2.500E-03 |

Then, we look at the table for the combination 1-and-0:

| | | | |
|---|---|---|---|
| R1 | 2.000E+04 | -1.250E-05 | -2.500E-03 |
| R2 | 2.000E+04 | 3.125E-06 | 6.250E-04 |
| R3 | 1.000E+04 | 1.250E-05 | 1.250E-03 |
| R4 | 2.000E+04 | 3.125E-06 | 6.250E-04 |
| Vmsb | 1.000E+00 | 5.000E-01 | 5.000E-03 |
| Vlsb | 0.000E+00 | 2.500E-01 | 0.000E+00 |

Finally, we look at the table for the combination 1-and-1:

| | | | |
|---|---|---|---|
| R1 | 2.000E+04 | -6.250E-06 | -1.250E-03 |
| R2 | 2.000E+04 | -4.688E-06 | -9.375E-04 |
| R3 | 1.000E+04 | 6.250E-06 | 6.250E-04 |
| R4 | 2.000E+04 | 7.813E-06 | 1.563E-03 |
| Vmsb | 1.000E+00 | 5.000E-01 | 5.000E-03 |
| Vlsb | 1.000E+00 | 2.500E-01 | 2.500E-03 |

Now we scan the tables for the resistors only, since we want to check the design against component variation. To calculate the normalized, worst-case deviation, we add the absolute value of the normalized deviations, for the resistors only, for each table. The absolute value is used because we assume that the resistors will deviate in the direction that changes the output voltage the most, à la Murphy's Law. The table with the worst deviation is the third table with a value of 5mV for each percent of resistor value change. This would suggest that for this digital-to-analog converter, a specification requiring 10mV maximum output deviation would allow the use of 2 percent (tolerance) resistors.

**Exercise 3.3.1**

Using the digital-to-analog converter example, now assume that the reference voltage (used at the inputs) has a 1 percent deviation. For each table, calculate the deviation due to the reference voltage, and subtract that result from a system specification of 10mV maximum deviation at the output. Using the remaining "allowable" deviation, what tolerance of resistors must be used to meet the system specification?

# CHAPTER 4

# DC Sweep

The simulations we have looked at so far calculated only quiescent, or DC, operation where the voltage or current sources maintained a fixed value. In this chapter we will look at circuits where the sources vary, though the analysis will still calculate quiescent (DC) operation. Using this type of analysis allows you to look at the results from many .OP analyses in a single simulation run. That is why it is called a "sweep."

Later in the chapter we will look at "controlled" sources. These allow you to build function blocks to transform signals.

## 4.1 SWEEPING A SOURCE

The DC sweep analysis is controlled with a ".DC" statement. When you "sweep" a source the simulator starts with one value for a source (voltage or current), calculates the DC bias-point (exactly as it does for the .OP analysis), then increments the value and does another DC bias-point calculation. This increment-then-analyze procedure continues until the last source value has been analyzed. You get to select the starting value, increment, and final value for the sweep. The results are the same as doing many .OP analyses, but is faster if you want to check the range of source values due to (i) new types of output that are available for this analysis, and (ii) the way the calculations are done.

The calculations for the .DC analysis are faster than the set of equivalent .OP analyses if only for the reason that PSpice does not have to reread the circuit file each time and then do the calculations. Beyond that, having arrived at the solution to the circuit for the initial source value, the solution for the next source value is assumed to be relatively close to the first solution. The first solution provides an estimate for the second solution. Then, having found the solution to the circuit for the first and second values, the solution for the third value is "guessed" by linear extrapolation of the first two solutions. This provides the estimate for the third solution. From then on, PSpice extrapolates from the previous two solutions for the next estimate solution (PSpice does not make use of more than the previous two solutions because it has been found that the time required to calculate higher-order extrapolations is not worthwhile for shorter solution times).

## 4.2 THE .DC STATEMENT

The .DC statement specifies the values used during the DC sweep. The statement says which source value is to be swept, the starting value, the amount to increment the value each step of the sweep, and at what value to quit the sweep. In the syntax shown,

.DC *<source name>* *<start value>* *<stop value>* *<incr. value>*

the *<source name>* is an independent source (voltage or current) in your circuit file. The .DC statement does not define the source, or how it is connected to the circuit. The .DC statement says only what values that source will have during the .DC analysis. You need to make sure that you have specified the source, in your circuit file, or PSpice will not be able to do the .DC analysis.

When adding a .DC statement to your circuit file, you do not need to change any of the other lines describing your circuit. Just add the .DC statement, as the sweep of values specified will override the fixed value indicated by the independent source statement (V source or I source) during the DC sweep analysis (only).

## 4.3 PRINTED OUTPUT

Being able to sweep through many values and calculate many results means you will want to get output that is different from the .OP analysis. Actually, it is the same output but the format has been changed so that it is more convenient to use. There are two things you would probably want to have in this new form of output: (i) to organize the calculated results so they could be referred to the value of the sweep source (like a table), and (ii) to be able to select which results are printed (to minimize the amount of output). The .PRINT statement does the job of selecting and tabulating results for the .DC analysis and other analyses to be explored later (and is used the same way for these other analyses, so learning it now will help later).

The .PRINT statement simply specifies which analysis the statement applies to, since it is used for many types of analyses, and which results to print. In the syntax shown, for printing results for the DC sweep,

.PRINT DC <*output value*> . . .

you can have many entries in the table of results, or output values. Each output value will get its own column in the table, and each row of the column will be the calculated result of the output value for each step in the DC sweep. The columns are in the same order as specified in the .PRINT statement. Usually you will want to print the sweep value in the first column to simplify finding results in the table, so PSpice does this for you; the first column, which comes before the columns you specify, always contains the value of the sweep variable.

The output values you can print are basically node voltages and device currents (which also means source currents, as a source is also a device). Node voltages can be printed relative to ground (node "0") or relative to another node (that is, the value printed is the difference of the voltages at two nodes). The syntax is shown for the DC sweep:

| | |
|---|---|
| .PRINT DC V(7) | to print the voltage at node 7 |
| .PRINT DC V(6,3) | to print (voltage at node 6) minus (voltage at node 3) |
| .PRINT DC V(R1) | to print the voltage across R1 (any two-terminal device) |

Device (and source) currents may be printed using the syntax:

| | |
|---|---|
| .PRINT DC I(R4) | to print the current through R4 (any two-terminal device) |

You can, as mentioned earlier, print several values in one table, and mix voltages and currents; for example:

.PRINT DC V(3) V(4) I(R2)

**Exercise 4.3.1**

Add a .DC statement to the (bridge) circuit and sweep the supply voltage from 1 to 2 volts in .1 volt increments. Also add a .PRINT statement to print values for the supply (sweep) voltage and the current through two of the resistors. Does the table produced by PSpice indicate that this circuit is linear?

The calculated value of current through a device, such as a resistor, means positive current, which will be flowing from the more positive voltage level to the more negative voltage level. The value printed may have the opposite sign (that is, a negative value instead of positive) from the one you were expecting. This depends on the order of the nodes when you specified, say, a resistor, in your circuit file; the syntax "R4 3 5 150" means a 150-ohm resistor between nodes 3 and 5. If PSpice finds that node 3 is more positive than node 5, then PSpice will calculate a positive (value) current through R4. You can think of current through the device as positive current flowing into the first node in the line specifying the device.

**Exercise 4.3.2**

Using the previous example's circuit file, swap the nodes of one of the resistors specified in the .PRINT statement. How did the output change? Does this mean the circuit works any differently from before?

## 4.4 PLOTTED OUTPUT

After a few simulations using the DC sweep analysis, you may notice that looking through the printed table is getting tedious. Wouldn't it be nice to have the computer graph the results? Well, PSpice will print graphs if you specify a .PLOT statement. Then you can use either .PRINT or .PLOT to look at the results, or both.

The .PLOT statement is nearly the same as .PRINT; you specify the type of analysis the plot is for, and which results you want plotted. The output values have the same form as for the .PRINT statement. If you include two .PLOT statements in your circuit file you should get two plots. In the syntax for the DC sweep

.PLOT DC <*output value*> . . . [<*min range*>, <*max range*>]

notice that .PLOT will let you set the range of the output axis. If you do not specify the range, PSpice will automatically calculate a range which includes all of the output values.

**Exercise 4.4.1**

Redo the previous exercise showing the use of the PRINT statement with DC sweeps, but use a PLOT statement instead.

## 4.5 LINEAR CONTROLLED SOURCES

The controlled sources are one of the most useful (and overlooked) features of PSpice (and this is true of many other non-SPICE simulators). Controlled sources measure voltage or current and use the measured value to control their output (also a voltage or current). The transformation allowed between input and output is a multidimensional polynomial. Both the number of dimensions (that is, number of measured inputs) and degree of each polynomial are set by the user (that is, you). But first, we will consider the linear case for these controlled sources before trying anything more difficult.

Allowing two types of input (voltage and current) and output yields four combinations of input/output:

the voltage-controlled voltage source (VCVS)
the current-controlled current source (CCCS)
the voltage-controlled current source (VCCS)
the current-controlled voltage source (CVVS)

These four sources are devices, just like resistors, and are given PSpice device types of E, F, G, and H, respectively. Mathematically, you may think of these devices as functions:

VCVS is the function $v_0 = E(v_1, v_2, v_3, \ldots)$
CCCS is the function $i_0 = F(i_1, i_2, i_3, \ldots)$
VCCS is the function $i_0 = G(v_1, v_2, v_3, \ldots)$
CCVS is the function $v_0 = H(i_1, i_2, i_3, \ldots)$

We specify these sources in a way that is similar to the passive devices we have been using: name, connecting nodes, and (instead of value) the transforming polynomial. As you might have guessed:

Exx is a voltage-controlled voltage source
Fxx is a current-controlled current source
Gxx is a voltage-controlled current source
Hxx is a current-controlled voltage source

In the most simple form, an example of a voltage-controlled voltage source,

```
E2 5 7 3 4 10
```

is a voltage source, whose output nodes are 5 and 7 (remember, the positive current is flowing out of the connection to node 5), and whose output voltage is controlled by the voltage present at nodes 3 and 4, with a simple multiplying gain of a factor of 10. This is the same as the equation

$$v_5 - v_7 = 10 \cdot (v_3 - v_4)$$

You may also want to include, for clarity, some superfluous (to PSpice) commas and parentheses to identify the input nodes. For example, the form

```
E2 5 7 (3,4) 10
```

makes the statement more device-like (a name, followed by two nodes, and then the value specification).

The same form may be used for the voltage-controlled current source, except that the multiplying value is converting voltage to current so it has the dimension of amps/volts. So instead of ''gain'' we have ''conductance,'' but as there is a transfer from one set of nodes to another set, it is called ''transconductance.''

**Exercise 4.5.1**

Write down a VCCS statement for the function

$$i_3 = (v_5 - v_7)/20\text{ohms}$$

What are the units for the transconductance value?

When the measured input is current, the syntax is different. PSpice needs to be told which current, that is, the current through which device. To simplify matters,

PSpice measures currents through voltage sources (the fixed value V devices, not the variable E or H devices described here). Instead of controlling nodes, the syntax includes the name of the V device that has the controlling current; for example,

F4 3 5 V2 5

is a current source whose output current is 5 times the current flowing through V2. Again, this is a simple amplification gain in current. If the device were an H device (CCVS) instead, there is a transformation from current to voltage. Then the units are volts/amps or "resistance," but again because of the transfer to another set of nodes, it is called "transresistance."

**Exercise 4.5.2**

Write down a CCVS statement for the function

$$v_3 - v_5 = I(V5) \cdot 20\text{ohms}$$

What are the units for the transresistance value?

You probably didn't realize that you have already used one of the controlled sources in the previous example circuits. To PSpice the resistor is simply a voltage-controlled current source, with the same input and output terminals! To the rest of a circuit

R5 3 2 120

is the same as

G5 3 2 (3,2) 120

The only difference is that when PSpice checks your circuit, the G device is a current source (with infinite impedance) and does not qualify as a DC path to ground.

**Example 4.5.3**

Review some of the previous exercises and replace resistors with G devices. Use .PRINT and/or .PLOT to verify that the operation of these circuits has not changed.

## 4.6 POLYNOMIAL CONTROLLED SOURCES

What is different about a polynomial controlled source (some call these "non-linear" sources), from the linear case, is how the polynomial is described to PSpice. First, the dimension of the polynomial is specified. Then, the inputs to be measured must be described. Finally, the coefficients of the polynomial are specified. Single-dimension polynomial functions, which are basically the additive combination of many linear functions, are easily described using the syntax

(*one of* E, F, G, H)xx <*node*> <*node*> POLY(1) <*inputs*> <*coefficients*>

As you may have guessed, polynomials in two dimensions (that is, having two controlling inputs) will use "POLY(2)" instead.

The list of coefficients are in order of ascending powers, that is, the list of coefficients a, b, c, etc., come from the formula

$$a + b \cdot x + c \cdot x^2 + d \cdot x^3 + \cdots$$

where every coefficient up to the last non-zero coefficient must be specified. For example, if you wanted the formula

$$1 + 2 \cdot x^3$$

you would specify the coefficient list

1 0 0 2

This way PSPice knows you are specifying a third-degree polynomial and that the coefficients of the higher degrees are all zero.

The controlling inputs come as pairs of nodes (for voltage inputs) or V device names (for current inputs). There must be as many pairs, or names, as there are dimensions to the polynomial. For now, we will focus on single-dimension polynomials.

**Exercise 4.6.1**

Create a circuit file that does a DC sweep from −2 to +2 volts in .1 volt increments. Add to the file a VCVS which implements the function $x^3 - x$. Use .PRINT and .PLOT to check the output of the controlled source. Try this again with a VCCS.

When you want to have an output that is the sum of other input functions, for example,

$$\text{output} = f_1(x_1) + f_2(x_2) + \cdots$$

you do not need to use a higher dimension POLY(); simply add the outputs. If they are voltage outputs, put the controlled sources in series. The voltages combine so that the voltage across the series is the sum of the individual voltages. The same is true for current outputs, except that you will want to have them in parallel with each other. This makes it easy to check the correctness of each function by itself before combining it with the other functions.

**Exercise 4.6.2**

Redo the previous exercise's function $x^3 - x$ as the sum of two functions: $x^3$ and $-x$. Run the DC sweep and check each function separately. Are there other ways to implement the $-x$ function?

## 4.7 GRAPHICS OUTPUT[1]

For those of you used to the world of computer graphics, the output from the simulator that we have looked at seems primitive. What is missing is the ability to look at the response of your circuit on the computer display. After all, even some pocket calculators will plot graphic functions. For PSpice, we have such a facility called "Probe." To use Probe you must first tell PSpice to create a data file for Probe, which is done by using the .PROBE statement.

The .PROBE statement is similar to the .PRINT and .PLOT statements. With .PROBE you may select node voltages and device currents to be output from the simulation. However, this is generally not the best use of the .PROBE statement; if you just put .PROBE in your circuit file, without specifying any particular outputs, PSPice will **save all** of the node voltages and device (branch) currents. Then later, when you are using Probe to look at the results of the simulation, everything has been saved for your inspection. You select the waveforms you want to view and, if you are curious about the operation of some section of your circuit, you may view those without rerunning the simulator (which is what you would need to do if the only output you had was .PRINT or .PLOT).

In the syntax shown, for saving results of any of the analyses,

.PROBE [*output value*] . . .

remember that if you want to save all of the voltages and currents from the simulation that you do not specify any output values. This "default" mode tells PSpice to save everything. If you do specify any output values, then only those that are specified will be saved. You would normally do this only to save room in the data file, which PSpice is making as input for Probe.

The file created by PSpice for use by Probe is called PROBE.DAT. This file has a structure that tells Probe what analyses, voltages, currents, and independent variables (such as time, frequency, etc.) are available. Of course, the output data is in the file, too. For most of the circuits you will simulate, even when saving all of the results, the PROBE.DAT files that are generated are quite modest in size by today's standards. Many complete circuit files, with the Probe data file output from PSpice, may be saved on one of the low-density IBM-PC diskettes. When you are saving a circuit, having a PROBE.DAT file for that circuit with all of the output variables saved is valuable if you ever want to review the simulation results.

However, since PSpice saves outputs at each step during the simulation, long simulations of large circuits will generate considerable output. Since you are usually interested in only a few voltages or currents, you can specify that PSpice save only those items to reduce the amount of data in the PROBE.DAT file. For large

[1] This section introducing graphics is placed here because it is helpful for you to be using waveforms graphics for the rest of the book. Also, this is the first opportunity we have to look at any interesting waveforms.

circuits, specifying a few output variables will also speed the simulation as PSpice will skip the operations done to save all of the other output variables.

Probe is simple to use and is menu driven so you don't need to remember any commands or statements (as with PSpice). Most of your difficulty will probably come from getting Probe set up and started the first time. Then it will be easy and you can focus on your simulations.

**Exercise 4.7.1**

Put a .PROBE statement in the circuit file for the sweep of the function $x^3 - x$. Use Probe to examine the output you had plotted using the .PLOT statement.

**Exercise 4.7.2**

What other ways could you generate or display the curve of $x^3 - x$?

## 4.8 MULTIPLE-INPUT CONTROLLED SOURCES

When the controlled source you want is a function of several inputs, describing the coefficients can get complicated. In general there are many possibilities and the general form must allow for all of them. PSpice is told only the number of dimensions (that is, the number of "measuring" inputs) for the polynomial, so the list of coefficients follows a rule to describe the function you want. The general form description is complex to follow, but we can look at a three-input case to get the pattern of the general form. Assuming our inputs are $v_1$, $v_2$, and $v_3$, and a list of coefficients called $k_0$, $k_1$, $k_2$, . . . , the polynomial form for three inputs is

| | |
|---|---|
| the constant term | $k_0 +$ |
| plus, the linear terms | $k_1 \cdot v_1 + k_2 \cdot v_2 + k_3 \cdot v_3 +$ |
| plus, cross terms | $k_4 \cdot v_1^2 + k_5 \cdot v_1 \cdot v_2 + k_6 \cdot v_1 \cdot v_3 +$ |
| | $k_7 \cdot v_2^2 + k_8 \cdot v_2 \cdot v_3 +$ |
| | $k_9 \cdot v_3^2 +$ |
| plus, more cross terms | $k_{10} \cdot v_1^3 + k_{11} \cdot v_1^2 \cdot v_2 + k_{12} \cdot v_1^2 \cdot v_3 +$ |
| | $k_{13} \cdot v_2^3 + k_{14} \cdot v_2^2 v_3 +$ |
| | $k_{15} \cdot v_3^3 + \cdots$ |

and so on. Obviously it will be easy to make errors if we have many inputs (which rarely happens, fortunately)!

The most basic use of the general case is to sum several input voltages. A four-input voltage summer would have the form

```
Eout 7 0 poly(4) (1,0) (2,0) (3,0) (4,0) 0 1 1 1 1
```

Again, notice the parentheses around the voltage node pairs, which are also comma separated. You may do this in PSpice, and most SPICEs, as the commas and parentheses are treated like spaces. This improves the "readability" of the polynomial form.

**Exercise 4.8.1**

Write down a voltage summer with the following weighted inputs:

$$1 \cdot v_1 + 3 \cdot v_2 + 2 \cdot v_3 + .5 \cdot v_4$$

Another common case is a two-input controlled source. You may have thought about how to multiply two voltages. Using the syntax described above, a voltage multiplier would be

E2 3 4 poly(2) (7,8) (5,6) 0 0 0 0 1

where the output voltage (across nodes 3 and 4) is a function of the input voltages (across nodes 7 and 8, and nodes 5 and 6). The offset coefficient (that is, the coefficient for the zeroeth degree) is zero, as well as the coefficients for linear voltage terms. That's three zeros, so far. The fourth zero is for the quadratic voltage term of the first input voltage. Then we arrive finally at the coefficient for the multiplication of the two input voltages, which is set to unity.

Arriving at a cubic function of two inputs is too painful using the general syntax. It is easier for you to decompose the function into two stages (if this is possible); for example,

$$xy^3$$

becomes

$$(x)(y^3)$$

This way you create the individual functions (and test them, if you are uncertain about their operation), and then combine the intermediate outputs with the multiplier described previously.

## 4.9 FUNCTION MODULES

By using the controlled sources you can create a variety of modular function blocks. With a set of these in your "simulator toolkit," you can quickly check circuit ideas. And, even though we are investigating these using DC sweep analysis, these blocks also work for all of the other types of analyses.

We have already covered the voltage-multiplier function using a controlled-source statement

E*name* <*+out node*> <*−out node*>

\+ POLY(2) <*+A node*> <*−A node*> <*+B node*> <*−B node*>

\+ 0 0 0 0 1

which will multiply the voltages across the node pairs A and B.

**Exercise 4.9.1**

Create a current multiplier which multiplies the current flowing through two independent voltage sources. Test it to be sure the direction of current output is what you expected.

If you were to connect the same nodes to both the A and B input of the voltage multiplier, the output would then be the square of the input voltage. This is a trivial extension of the multiplier. However, if we include the multiplier in a feedback loop, we can develop new uses.

Feedback theory tells us that a circuit, as shown in Figure 4.1, which has three major sections—(i) a forward path, including a perfect (one hopes) amplifier, (ii) a feedback path, which has the interesting circuitry, and (iii) a difference block, which creates the "error" signal—will develop the transfer function

$$\text{output} = \text{input} \cdot A/(A \cdot K + 1)$$

so that the gain of the circuit is

$$\text{gain} = \text{output/input} = A/(A \cdot K + 1) = (1/K) \cdot A \cdot K/(A \cdot K + 1)$$

so that if the forward amplification gain, A, is large, the circuit gain becomes

$$\text{gain} = 1/K$$

In PSpice, we can easily create "perfect amplifiers" that are linear and have huge gain ratios, with the controlled-source statement. In fact, we can even integrate the amplifier with the difference function to create a "perfect error amplifier" by implementing the function

$$output = gain \cdot input - gain \cdot feedback$$

Now for the feedback section. If the entire circuit were to perform, say, the square-root function, then the function the feedback section has to perform is

$$gain = output/input = input^{1/2}/input = 1/K$$

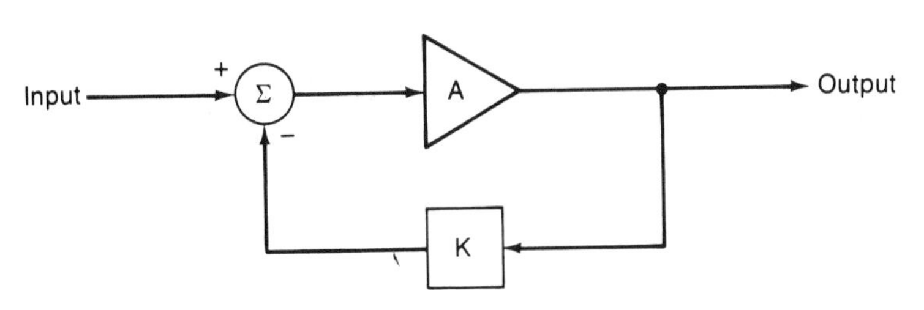


**Figure 4.1** Schematic of generalized feedback circuit.

This means the feedback section has to square the output voltage, a function we just covered earlier. Now we can build a circuit that calculates, with a small degree of error, the square-root of the input voltage.

```
Square-root circuit
Vin 1 0 0
Rin 1 0 1E6
Efwd 2 0 poly(2) (1,0) (3,0) 0 1E6-1E6 ; error amplifier
Rfwd 2 0 1E6
Erev 3 0 poly(2) (2,0) (2,0) 0 0 0 0 1 ; feedback section
Rrev 3 0 1E6
.DC Vin 0 10 .1
.PROBE
.END
```

Notice that resistors were placed across all of the source outputs. Without them the source outputs would be dangling, for the inputs to the controlled sources are considered by PSpice to have infinite impedance.

**Exercise 4.9.2**

Build and run the square-root circuit. Check the output values. How could the feedback section be simplified to have only one input? Try a different DC sweep, starting at +5 volts and sweep to −5 volts. Why doesn't the circuit work for negative input voltages?

**Exercise 4.9.3**

Build and test a cube-root circuit. Does it work for negative input voltages?

Now suppose we wanted a circuit whose output would be the ratio of the two input voltages. Then the function the feedback section has to perform is

$$gain = output/input1 = (input1/input2)/input1 = 1/input2 = 1/K$$

This means the feedback section has to multiply the output voltage by the denominator point. So now we can build a circuit that calculates, with a small degree of error, the ratio of two input voltages.

```
Divider circuit
Vtop 1 0 1 ; top of fraction
Rtop 1 0 1E6
Vbot 2 0 1 ; bottom of fraction
Rbot 2 0 1E6
Efwd 3 0 poly(2) (1,0) (4,0) 0 1E6-1E6 ; error amplifier
Rfwd 3 0 1E6
Erev 4 0 poly(2) (3,0) (2,0) 0 0 0 0 1 ; feedback section
Rrev 4 ø 1E6
.DC Vbot .1 1 .05
.PROBE
.END
```

**Exercise 4.9.4**

Build and test the divider circuit. Try extending the function to be the ratio of two independent, second-order polynomials.

## 4.10 SUBCIRCUITS

Now that we have built some useful function blocks, there is still the problem that for each use of a block we would have to rekey the section of circuit into our circuit file. Furthermore, if there were multiple blocks of the same type, unique nodes and component names would have to be used. This quickly becomes tedious and inflexible. To help out, PSpice has a macro facility called a ''subcircuit'' which captures a circuit function as a ''subnet'' of connected components. Terminals are assigned for the subcircuit to connect it into your circuit. The subcircuit definition has the form

.SUBCKT <*definition name*> <*node* 1> <*node* 2> · · ·

*components defining circuit function*

.ENDS

where the list of nodes identify which of the nodes of the subcircuit are terminals that may be attached to the external circuit. Node numbers within the subcircuit are separate from, and are not to be confused with, any nodes that might be in the external circuit (with the exception of the ''0'', or ground, node, which is a global node reference). For example, we could encapsulate the square-root function from before by the following subcircuit definition:

```
.SUBCKT SqRoot 1 2 3 4
Efwd 3 4 poly(2) (1,2) (5,0) 0 1E6 -1E6; error amplifier
Erev 5 0 poly(1) (3,4) 0 0 1 ; simplified feedback section
Rrev 5 0 1E6
.ENDS
```

Notice that the input nodes (1 and 2) and the output nodes (3 and 4) are ''floating,'' so that the function block is now similar to the controlled-source statements and may be inserted in any circuitry.

To use the subcircuit in a circuit file PSpice uses the convention that these ''new'' devices are treated as a new device type, whose names start with an ''X,'' as follows:

X*name* <*node* 1> <*node* 2> · · · <*subcircuit definition name*>

Each of the terminal nodes in the subcircuit definition must be used when the subcircuit is used, and the order of the nodes is the same as the order used in the definition of the subcircuit. For example, we will rewrite the test circuit for the square-root function from before:

```
Square-root test circuit
Vin 1 0 0
Rin 1 0 1E6
Xblock 1 0 2 0 SqRoot ; here is where we use the subcircuit
Rout 2 0 1E6
*Square root definition
.SUBCKT SqRoot 1 2 3 4
Efwd 3 4 poly(2) (1,2) (5,0) 0 1E6 -1E6 ; error amplifier
Erev 5 0 poly(1) (3,4) 0 0 1 ; simplified feedback section
Rrev 5 0 1E6
.ENDS
*
.DC Vin 0 10 .1
.PROBE
.END
```

The subcircuit definition may be anywhere between the title line and the .END statement. With PSpice you may collect all of these function blocks into a "tool kit" library file that will be searched by the command.

.LIB <*filename*>

to bring in only the subcircuit definitions needed by a particular circuit file. You can use these library files to save and reuse function blocks that are useful to your work.

# CHAPTER 5

# Transfer Function

The "transfer function" analysis is another DC-bias analysis and is used to calculate some external, or "black box," characteristics of your circuit. The output of this analysis are values for (i) "small-signal" DC gain (input-to-output transfer ratio), (ii) DC input resistance, and (iii) DC output resistance. We will examine what these values tell you about your circuit.

## 5.1 SMALL-SIGNAL DC ANALYSIS

The transfer function of your circuit involves "small-signal" DC analysis. "Large" signals are the normal excursions that your circuit might encounter; for example, the output fluctuations of a stereo amplifier are generally large. "Small" signals are minuscule; for example, the signal amplified by a radio receiver is quite small. However, "small-signal" analysis deals with circuit operation in the limit of signals approaching zero strength. When your circuit has linear operation for the signals it will normally encounter, then small-signal calculations may be applied to an operation that is merely "small-ish." Large signals need to be treated in a different fashion, which we will get to in due course.

Even though small-signal analysis is for small (nearly zero level) signals, this does not mean that your circuit is "turned off" and all the nodes are at zero volts. It just means that the stimulus to the circuit is small, but the power supply, for example, has its normal value. The situation where your circuit is energized by DC sources, but there is no other external stimulus, is the bias-point situation we have looked at before. Small-signal DC analysis, then, does calculations of the effects of minuscule input stimulation to your circuit. It answers, for example, the question, "If the input node were to deviate slightly from its bias-point value, what would the output do?"

## 5.2 CIRCUIT GAIN

The most common question about the small-signal operation of a circuit is, "What is the gain?" This is probably because the most common electronic circuit is an amplifier (even digital circuits amplify), and its gain is a fundamental specification. Gain is the ratio of output signal deviation to input signal deviation. We use the word "deviation" because we need to differentiate between the quiescent, or steady-state, level of the input and the small excursions that represent the "real" signal. Mathematically, small-signal DC gain is the derivative of output with respect to input, at the DC bias-point (and at zero frequency). For example,

$$dV_{out}/dV_{in}$$

is an expression for voltage gain. There are other types of "gain" you may be interested in. As you might have guessed,

$$dI_{out}/dI_{in}$$

is an expression for current gain. However, you may also want to evaluate

$$dI_{load}/dV_{in}$$

which is an expression for transconductance, or

$$dV_{out}/dI_{in}$$

which is an expression for transresistance.

## 5.3 INPUT AND OUTPUT RESISTANCE

Resistance, as you will recall from Ohm's law, is the ratio of voltage across the resistor to current flowing through the resistor: $V = I \cdot R$ or $R = V/I$. The input and output resistance of a circuit is much the same, although we are now referring to the "dynamic," or small-signal, resistance at the input or output. Mathematically, small-signal DC resistance is the derivative of the input voltage with respect to the input current, at the DC bias-point (and at zero frequency). For example,

$$dV_{in}/dI_{in}$$

is an expression for input resistance.

## 5.4 THE .TF STATEMENT

The ".TF" statement specifies what you consider to be the "input" and "output" of your circuit (PSpice doesn't know how your circuit is being used). Once the DC bias-point for the circuit is calculated, PSpice calculates the following "black box," or "transfer" functions: gain, input resistance, and output resistance. The format for the statement is

.TF *<output variable name>* *<input source name>*

where the *<output variable name>* is in the same format as for the .PRINT statement. The *<input source name>* must be an independent source (V or I device); this is because the input usually has some fixed input bias, even if it is zero, which you may want to set (also, remember that PSpice will not analyze circuits with "dangling" nodes).

## 5.5 TRANSFER FUNCTION ANALYSIS

Having a .TF statement in your circuit file causes the transfer function calculations to be done when the DC bias-point calculations are completed. You do not need to specify any other output, such as .PRINT, to get the results of the transfer function analysis. We can now try working some examples to see what .TF will do. Usually you will be using .TF to analyze a more complicated, active device circuit, such as a transistor amplifier. However, for ease of understanding we can demonstrate the use of .TF with linear and nonlinear controlled sources as our active elements.

Consider the simple circuit

```
Resistor divider
Vin 1 0 1volt
R1  1 2 3ohm
R2  2 0 1ohm
.TF V(2) Vin
.END
```

which we will analyze. PSpice is run, and in the output we will find (something similar to):

```
V(2)/Vin                    = 2.500E-01
INPUT RESISTANCE AT Vin     = 4.000E+00
OUTPUT RESISTANCE AT V(2)   = 7.500E-01
```

What does this mean? Since we specified both input and output variables as a voltage at a node and a voltage source, respectively, the transfer function gain calculation is voltage gain. Our resistor divider has a voltage gain of ¼, meaning that the variation of the voltage at V(2) is one-quarter of the variation in VIN. The input resistance "seen" by (input) VIN is 4 ohms, as you might have expected. This means that a variation in the voltage (in volts) of VIN will be four times the measured current variation (in amps) of VIN caused by the voltage variation. The output resistance "seen" by (output) node 2 is ¾ ohms. This means that if you could manage to vary the voltage at node 2, the size of the variation (in volts) would be three-quarters the size of the current (in amps) required from the means by which you accomplished the variation. How can we check these calculations?

Checking the value for gain is fairly simple; the .TF statement also caused the output of the bias-point calculation. We can see that the ratio of V(2) to VIN is ¼; the circuit has only linear elements, so doubling the value of VIN will double the value of V(2).

**Exercise 5.5.1**

> Run the transfer function analysis just described. Suppose you don't immediately see that doubling the value of VIN will double the value of V(2) for the DC bias-point . . . Try it.

Checking the input resistance "seen" by VIN is elementary, too. The output from PSpice also included the current supplied by VIN. Since VIN is the only source of current to the circuit, we may divide VIN's voltage by its supply current to arrive at the resistance "seen" by VIN. You might be concerned that the literal answer to this formula is −4 (ohms), however the −0.25 amps of current in the output file indicates just that the current is flowing out of VIN's first node. This is from the SPICE convention for current direction: positive currents flowing into a terminal have a positive value.

Checking output resistance is easy, too. Since it was stated that output resistance was the ratio of the output voltage change to an external influence's current deviation, let us try that. By adding the current source

```
IOUT 2 0 0.1amp
```

and resimulating, we can "draw down" the output voltage. Now we see that the voltage at V(2) is now 0.175 volts, instead of the 0.25 volts it was before. The output resistance is calculated as

$$(0.25 - 0.175)/0.1 = 0.75$$

From the perspective of node 2, R1 and R2 are in parallel because the voltage source VIN is ideal and has zero resistance. You could have calculated the output resistance as the parallel combination of R1 and R2

$$1/(1/R1 + 1/R2) = 1/(1/3 + 1/1) = 3/(1 + 3) = 3/4 = 0.75$$

**Exercise 5.5.2**

Try the method just described to check the output resistance. Now, instead of using a current source, use a voltage source to set the voltage of V(2) to 0.3 volts. How much current was supplied by the new voltage source? How can you use this information to calculate output resistance? Did you get the same value?

## 5.6 LINEAR EXAMPLE

What kind of transfer function information is calculated for a circuit with linear gain? Consider the circuit in Figure 5.1, represented by the following circuit file:

```
Simple gain-of-5 circuit
Vin  1 0 1volt
Rin  1 0 1ohm
Gout 0 2 1 0 5.0
Rout 2 0 1ohm
.TF V(2) Vin
.END
```

This circuit is an ideal ''gain block'' with input and output resistance (both are 1 ohm). ''Gout'' was connected so that positive current flows into node 2, making the gain of the circuit a positive value; when V(1) increases in value, then V(2) will also increase in value. The result of the simulation will be something similar to

```
V(2)/Vin                    = 5.000E+00
INPUT RESISTANCE AT Vin     = 1.000E+00
OUTPUT RESISTANCE AT V(2)   = 1.000E+00
```

These were the results we expected.

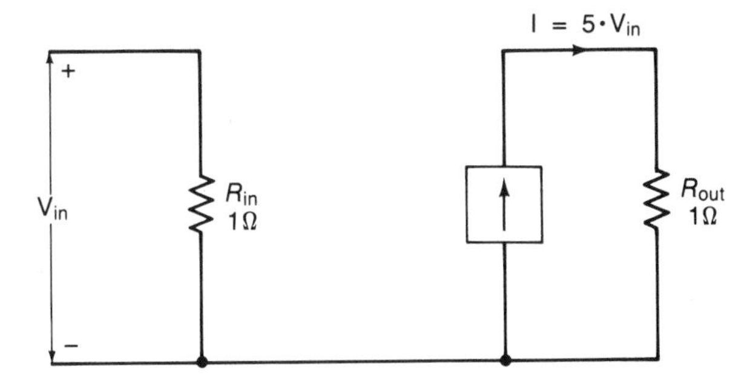


**Figure 5.1** Schematic for gain-of-5 circuit example.

**Exercise 5.6.1**

You may have noticed that, in the previous example, Vin was set at 1 volt. Does this matter? Try setting it to a different value, say 2 volts, and rerun the simulation. What changed? Why didn't the output of the transfer function analysis change?

**Exercise 5.6.2**

Assuming you know about Thevenin equivalent transformations, change the example above to use an "E" device as the gain element. Check your work by running the simulator to see that the same values are calculated for the transfer function analysis. (Hint: Your output resistance may not "dangle," so connect the output node to ground with a large resistance.)

## 5.7 NONLINEAR EXAMPLE

Electronic circuitry is nonlinear, however the transfer function analysis does linear calculations on the circuit. That is, the calculations are done once the circuit has attained its bias-point and has been "linearized." This means that all of the elements in the circuit are expressed as their linear equivalents, which are valid only for that particular bias-point. A different bias-point would probably require a different "linearization." So far we have looked only at examples that were already linear, so the process of linearizing these elements did not change them. Consider the following example, which has a nonlinear element:

```
Nonlinear gain circuit
Vin  1 0 1volt
Rin  1 0 1ohm
Gout 0 2 poly(1) (1,0) 0 0 1
Rout 2 0 1ohm
.TF V(2) Vin
.END
```

This circuit is very much like the previous example, which was linear. However, the gain element is nonlinear as its output is the square of the voltage at node 1. After running the simulator, the output will be something similar to

```
V(2)/Vin                   = 2.000E+00
INPUT RESISTANCE AT Vin    = 1.000E+00
OUTPUT RESISTANCE AT V(2)  = 1.000E+00
```

The input is 1 volt, and the output is 1 volt (the input voltage squared), but why is the gain calculated as a value of 2.0? This is due to the linearization of the circuit at the bias-point. The linear slope value of the gain circuit is the mathematical derivative of the gain function. Remember, we defined small-signal voltage gain as

$$dV_{out}/dV_{in}$$

and now you can see the result of this definition. How can we check this value for gain? Rerun the simulation with Vin set to 1.01 volts. This time the values printed for the bias-point will show V(2) as 1.0201 volts, so the gain of the circuit for small excursions in the input voltage is

$$(\text{change in V(2)})/(\text{change in Vin}) = (1.0201 - 1)/(1.01 - 1) = 2.01$$

You can see that, for small deviations in the input voltage, the output deviations are twice as large. In this case, PSpice shows enough precision in the printed values to see the effect of the square function; we can relate the output from the last two simulation to what we know about "Taylor series" from mathematics.

Taylor series involve the mathematical linearization of functions, and the series allows you to calculate values for the function in the neighborhood of a point on the function (which we call the "bias-point" in electronics). If $f(x)$ has continuous derivatives in the region of a point $x = a$, then

$$f(x) = f(a) + f'(a)(x - a)/1! + f''(a)(x - a)^2/2! + \cdots$$

From the first simulation of the "voltage squared" we have the result for $a$ = 1.0000 and $f(a)$ = 1.0000; these are the bias-point node voltages. Now, to predict the results of the second simulation, with $x = 1.01$, we know

the original function: $f(x) = x^2$
its first derivative: $f'(x) = 2x$
its second derivative: $f''(x) = 2$

so the Taylor series calculation for $a = 1.00$ and $x = 1.01$ is

$$1.00 + 2 \cdot 1.00 \cdot (1.01 - 1.00)/1 + 2 \cdot (1.01 - 1.00)^2/2$$
$$= 1 + 0.02 + 0.0001 = 1.0201$$

which is the same result PSpice calculated for V(2) in the second simulation.

**Exercise 5.7.1**

The circuit above is a "voltage squared" function. What transfer function results would you expect if Vin were −1.0 volts? What bias-point would you expect if Vin were −1.01 volts? Check this with the simulator and using the Taylor series formula.

**Exercise 5.7.2**

Try similar simulations for the function $f(x) = x^3$.

## 5.8 PLOTTING SMALL-SIGNAL GAIN

Probe graphics can be used to plot gain over a sweep of DC operating conditions. Running several simulations with .TF analysis is tedious. Fortunately this technique does not use the .TF at all, but is included here as an alternate way to get small-signal gain numbers from PSpice. Note that this is for gain only. While there is an equivalent technique for sweep results of input or output resistance, it is much more interesting to look at sweeps of frequency, so we will look at this later in the book.

The definition of small-signal voltage gain is

$$dV_{out}/dV_{in}$$

Any of the variables or formulas you enter into Probe may be differentiated by wrapping them in the ''d()'' function. Also, single voltage or current values may be differentiated by using a ''d'' prefix. For example, the derivative of V(2) is dV(2). Probe is able to calculate approximations for derivatives by using divided differences. This follows the notion that in the limit, as $a$ approaches $x$

$$f(x - a)/(x - a) = f'(x)$$

If we are careful to make the differences small enough, then the calculations should be useful (we also should not make the differences too small, or Probe could lose accuracy in its calculations).

Using the ''voltage squared'' circuit from earlier in this chapter, we can look at its small-signal gain over a range of bias combinations using the DC sweep and Probe. The circuit would be set up as shown:

```
Nonlinear gain circuit
Vin  1 0 1volt
Rin  1 0 1ohm
Gout 0 2 poly(1) (1,0) 0 0 1
Rout 2 0 1ohm
.DC Vin -2 2 .05
.PROBE
.END
```

After the simulation is finished, start Probe (you may have this set up to happen automatically). The divided differences mentioned above are calculated relative to the displayed X-axis. For example, if Vin is the X-axis, then the Probe function dV(2) is shorthand for d(V(2))/d(Vin). By displaying the variables V(2) and dV(2), we see the plot of small-signal gain shown in Figure 5.2, which indicates the output voltage and the small-signal gain (approximately) for the circuit.

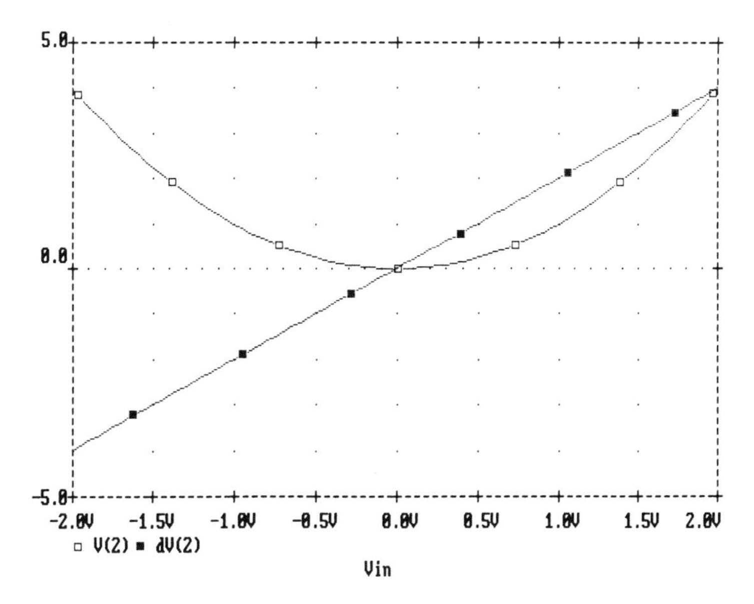


**Figure 5.2** Plot of small-signal gain.

### Exercise 5.8.1

Try setting the step size of the DC sweep to a larger value, say .2 volts, and see how the graph of V(2) and d(V2) changes.

### Exercise 5.8.2

Try the previous example and exercise with Gout as a cubic function.

# CHAPTER 6

# Frequency Response

One of the more popular uses of circuit simulators is to verify the frequency response of signal filter and control circuits. The frequency response analysis calculates all of the AC node voltages and branch currents over a swept range of frequencies. The output of this analysis are the values for (i) the amplitude of node voltages and device currents, and (ii) the relative phase angles of the node voltages and device currents.

By ''frequency response'' we actually mean ''small-signal frequency response,'' where the analysis is done with the assumption that the input signals are small enough to minimize nonlinear effects. Novice users of PSpice often confuse frequency response with ''transient response'' (which we will look at later). They relate frequency response to a lab-bench setup that sweeps the input of their circuit with an oscillator, while they look at the output with an oscilloscope. While it is true the oscillator is providing a waveform with a frequency, often the signal is large enough to induce nonlinear behavior in the circuit. This type of experiment may be simulated also, but not by using frequency response analysis.

At the beginning of frequency response (AC) analysis, the DC bias-point for the simulated circuit is calculated. PSpice uses the same procedure as for the DC bias-point (OP) analysis. Then the linear component equivalent of the circuit is ''saved'' (this is what is meant by saying the circuit was linearized) and used for the AC analysis. The laws of Ohm and Kirchhoff apply for AC analysis, too, but the impedances between nodes are said to be ''complex'' (having both a ''real'' and ''imaginary'' component) giving different results from DC analysis.

## 6.1 SPECIFYING INPUT SOURCES

From before, you recall that the independent voltage sources (V device) and current sources (I device) had the statement form

*<name> <node> <node> <value>*

where value was the DC voltage or current level, depending on the device type. Actually, the value part of the statement could be stated in the form

DC *<value>*

to remind you that the value is a DC bias amount, however PSpice knows that if you include only the numeric value it is implied that this is the DC value. To reiterate, the statement

VIN 1 0 0.5

is shorthand for

Vin 1 0 DC 0.5

and both represent the same voltage source.

We cover this because it may only be done for the DC value, but **not** the AC value. PSpice can allow "shorthand" for only one of the specifications. A fuller representation of the input source statement is

*<name> <node> <node> <DC value> <AC value>*

where if you leave out the DC value, the DC value is set to zero. Likewise for the AC value; if you leave it off, the AC value is zero. You may even leave off both values, so that

```
VIN 1 0
```

is shorthand for

```
Vin 1 0 DC 0.0 AC 0.0 0.0
```

and both represent a "zero volt" source (which we have used before as a current monitor). You will include values for both for situations where you want to use an independent source that has both a DC and an AC value (the AC signal rides on a DC input level).

The *<AC value>* portion of the statement has the form

AC *<magnitude value> <phase value>*

where you may leave off the phase value if the phase is zero. The *<magnitude value>* is straightforward and is simply the peak amplitude of the AC excitation. The *<phase value>* is the offset phase you want to have for this source, and the offset amount is relative to "zero phase." This may seem excessively general for most circuits where the output phase is already relative to the input phase, so there is usually no need to shift the input phase. However, PSpice can deal with multiple input sources of differing magnitudes and relative phases, so you may need to shift phase for a more complicated circuit.

Note that, since the circuit has already been linearized for this analysis, the excitation level you choose is arbitrary as the levels calculated for the rest of the circuit will change in proportion to the input. For this reason we normally set the input magnitude to unity, so that all of the calculated levels represent "gain."

## 6.2 THE .AC STATEMENT

The ".AC" statement specifies the frequency values used during the frequency response analysis. The statement says only which frequencies are used and **all of the independent sources' AC values will be set to these frequencies.** Usually there is only one source that has a non-zero AC value, and it becomes the input source of AC signal, but PSpice is not limited to just one AC signal source.

The frequency sweep comes in three types: linear, octave, and decade. Their syntax forms are similar. The statement

.AC LIN *<points value>* *<begin value>* *<end value>*

defines a linear frequency sweep, with *<points value>* specifying the number of points in the sweep starting at *<begin value>* frequency and finishing at *<end value>* frequency. The statement

.AC OCT *<points value>* *<begin value>* *<end value>*

defines a logarithmic frequency sweep, with *<points value>* specifying the number of points per octave (a twofold increase in frequency) in the sweep starting at *<begin value>* frequency and finishing at *<end value>* frequency. The statement

.AC DEC *<points value>* *<begin value>* *<end value>*

defines a logarithmic frequency sweep, with *<points value>* specifying the number of points per decade (a tenfold increase in frequency) in the sweep starting at *<begin value>* frequency and finishing at *<end value>* frequency.

## 6.3 PRINT AND PLOT OUTPUT

Output from AC analysis may be generated by .PRINT or .PLOT statements, just as in DC analysis. In either case the output is organized by the frequency at which the calculations were made. The statement forms are

.PRINT AC *<output value>* . . .

and

.PLOT AC *<output value>* . . .

Each *<output value>* entry becomes a column in the table output by the .PRINT statement, or a curve in the plot output by the .PLOT statement. The output values you can print/plot are node voltages and device currents (which also means source currents, as a source is also a device) with some special considerations for AC analysis. Node voltages and device currents may be specified as magnitude, phase, real, or imaginary, plus some functions, by adding a suffix to "V" (voltage magnitude) or "I" (current magnitude):

| | |
|---|---|
| (no suffix) | for magnitude |
| M | for magnitude |
| DB | for magnitude in decibels: 20·*log*(value) |
| P | for phase |
| G | for group delay (not in SPICE) |
| R | for real part |
| I | for imaginary part |

So, for example:

| | |
|---|---|
| .PRINT AC V(7) | prints the voltage magnitude at node 7 |
| .PRINT AC VP(7) | prints the voltage phase at node 7 |
| .PRINT AC IR(R1) | prints the real part of the current through R1 |

You may print several values in one table, and mix voltages and currents; for example,

.PRINT AC V(3) VP(4) II(R2)

Usually you will want to print the analysis frequency in the first column to simplify finding results in the table, so PSpice does this for you; the first column, which comes before the columns you specify, always contains the value of the analysis frequency.

## 6.4 GRAPHICS: BODE PLOTS

Using Probe with AC analysis is identical to what we did before with the DC sweep; just include a .PROBE statement to the circuit file. Let's try a small filter circuit to explore what kind of results we can get out of the simulator.

This circuit is a double-pole, low-pass LC-filter circuit. The only unusual feature about this circuit is that we have split the input resistance into several sections.

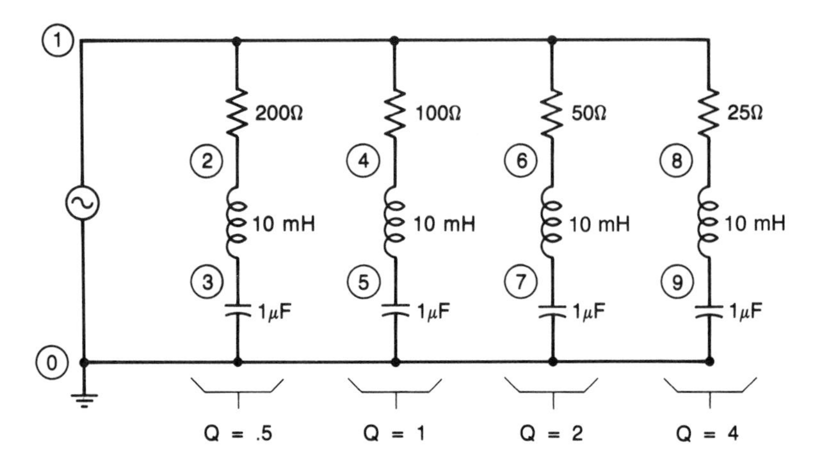


**Figure 6.1** Schematic of double-pole LC circuit example.

This will allow us to investigate the response of the circuit with several values of input resistance, all in one simulation. The equivalent circuit file is

```
Four double-pole, low-pass, LC-filters
VIN 1 0 AC 1
* Q = .5
R1 1 2 200
L1 2 3 10mH
C1 3 0 1uF
* Q = 1
R2 1 4 100
L2 4 5 10mH
C2 5 0 1uF
* Q = 2
R3 1 6 50
L3 6 7 10mH
C3 7 0 1uF
* Q = 4
R4 1 8 25
L4 8 9 10mH
C4 9 0 1uF
.AC DEC 100 100hz 10Khz
.PROBE
.END
```

Note that the input level selected for the AC source is 1 volt. Since frequency response analysis is a small-signal analysis, this simplifies looking at the ratio of output response to input response. If the input equals 1, there is no need to literally calculate the ratio since the output value is the ratio.

The circuit values were selected to have a resonant frequency in the audio range and different quality factor "Q" for each circuit section. Displaying our waveforms:

| | |
|---|---|
| vm(3) | is the magnitude response for $Q = \frac{1}{2}$ |
| vm(5) | is the magnitude response for $Q = 1$ |
| vm(7) | is the magnitude response for $Q = 2$ |
| vm(9) | is the magnitude response for $Q = 4$ |

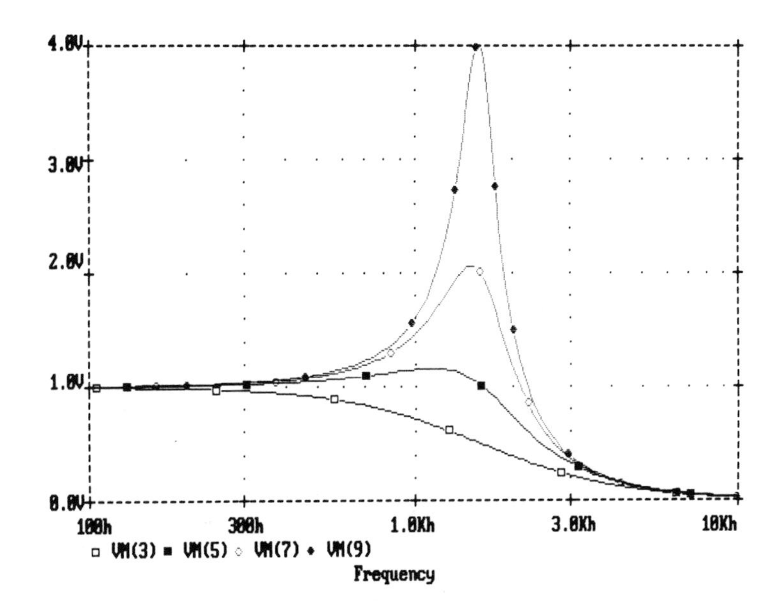


**Figure 6.2** Plot of magnitude response.

We may display phases to show the phase response of the filter for different values of Q:

| | |
|---|---|
| vp(3) | is the phase response for $Q = \frac{1}{2}$ |
| vp(5) | is the phase response for $Q = 1$ |
| vp(7) | is the phase response for $Q = 2$ |
| vp(9) | is the phase response for $Q = 4$ |

Since response is the ratio of output to input, you can look at the individual responses of a complicated filter chain this way. Probe will do the division for you, which is a function you rarely find on an oscilloscope. Furthermore, if you were displaying response in decibels, then you need only to subtract decibel level of the input from the decibel level of the output. For example, if you had a two-stage filter with stage "A" feeding stage "B," then the response of stage "B" is

$$db(v(B_0)) - db(v(A_0))$$

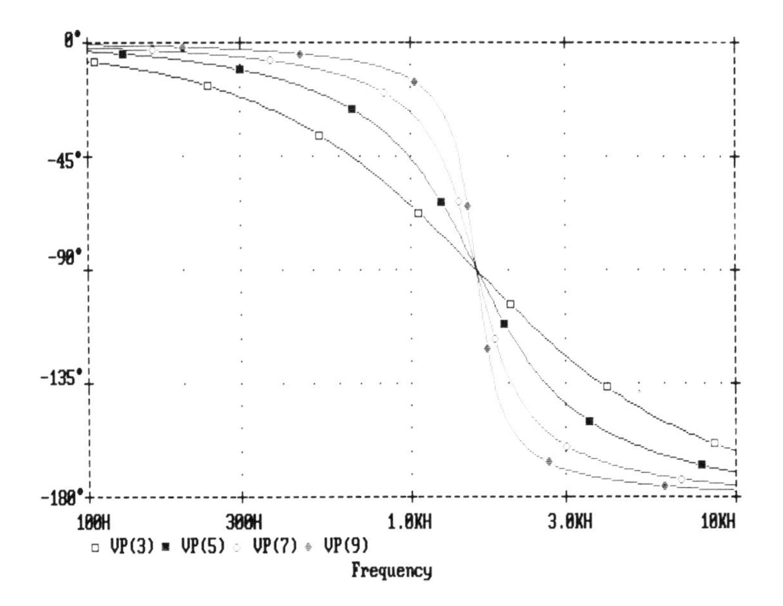


**Figure 6.3** Plot of phase response.

A similar technique works for displaying phase angles. In this case you merely subtract the input phase angle from the output phase angle. Using the previous example of the two-stage filter, the phase response of stage "B" is

$$\text{vp}(B_0) - \text{vp}(A_0)$$

## 6.5 PLOTTING GROUP DELAY

An important characteristic of a filter, or signal processing circuit, is its phase response, which is related to the distortion of the signal's waveform (shape) as it passes through the filter. (The phase response of the filter also demonstrates a link between frequency analysis and transient [time domain] analysis.) A circuit can maintain the wave-shape of a signal if the phase shift of each frequency component of the spectrum of the signal is a linear function of frequency. This means the circuit has uniform, or flat, delay; the signal is shifted only in time but is otherwise unchanged.

If the phase shift is a linear function of frequency, then its derivative with respect to frequency should be constant. As it happens, the formula we just described

$$-d\text{phase}/d\text{frequency}$$

is also the delay time through the circuit for each frequency component. In the realm of modulated transmission, this is the delay time of the components in the envelope of a modulated carrier, so the delay is called "envelope delay." In the realm of data transmission where the signals are pulses, this is the delay time for groups of pulses, so the delay is called "group delay." In either case the integrity of the transmission may depend on the variation in the delay time.

Probe will calculate group delay by using phase and frequency differences. So long as the group delay is not changing rapidly between each frequency increment in the analysis the error of this technique will remain small. We may display group delay to show the response of the filter for different values of Q:

vg(3) is the group delay for $Q = \frac{1}{2}$
vg(5) is the group delay for $Q = 1$
vg(7) is the group delay for $Q = 2$
vg(9) is the group delay for $Q = 4$

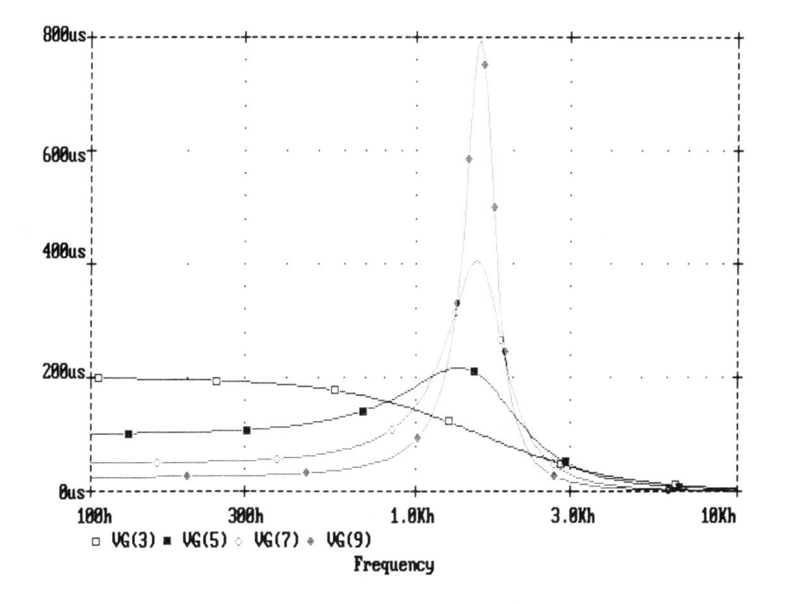


**Figure 6.4** Plot of group delay.

**Exercise 6.5.1**

Develop the formula that Probe uses for calculating group delay for output voltage. (Hint: use the derivative of the phase of the output voltage.) Display both plots to check your formula.

## 6.6 COMPLEX VALUES

When using frequency analysis, all of the calculated voltages and currents are complex values. That is, these quantities are expressed as complex numbers, which have both a **real** and **imaginary** component, not just scalar value. You may treat (mathematically) the values in Probe the same way you would treat complex numbers (some of those things you learned in math class might actually be useful, after all). Let's review the basics of manipulating complex values.

Suppose we have a complex number, "A," which is a combination of two real numbers, "a" and "b," by the formula

$$A = a + b \cdot i$$

where "i" is an imaginary quantity (equal to the square root of $-1$). The absolute value, or "modulus" (magnitude), of A is

$$|A| = (a^2 + b^2)^{1/2}$$

and the angle (phase) of A is

$$\angle A = tan^{-1}(b/a)$$

Using the magnitude and phase of A, we may respecify a complex value as

$$A = a + b \cdot i = |A| e^{i \angle A}$$

where $e$ is Euler's number (approximately 2.71828). This is another transformation available in complex math that we will take advantage of.

So, in Probe, when you call for the magnitude, or phase, of a voltage you are displaying the absolute value and angle of the complex quantity.

VM(5) is the magnitude of the voltage at node 5
V(5) is also the magnitude of the voltage, for convenience
VP(5) is the phase of the voltage at node 5

If we have two complex values

$$A = a + b \cdot i$$

$$B = c + d \cdot i$$

you may remember that the addition formula for complex numbers is

$$A + B = (a + b \cdot i) + (c + d \cdot i) = (a + c) + (b + d) \cdot i$$

This is the technique Probe uses when you add, or subtract, complex voltages or currents.

The multiplication formula is

$$A \cdot B = (a \cdot c - b \cdot d) + (b \cdot c + a \cdot d) \cdot i = |A| \cdot |B| \cdot e^{i(\angle A + \angle B)}$$

and the division formula is

$$A/B = (a \cdot c + b \cdot d)/(c^2 + d^2) + (b \cdot c - a \cdot d) \cdot i/(c^2 + d^2) = |A|/|B| \cdot e^{i(\angle A - \angle B)}$$

These techniques are also in Probe for multiplying and dividing complex voltages or currents. All of these formulas become handy when you want the equations like $V = I \cdot R$, but V and I are complex quantities. We'll try this in the next section.

## 6.7 PLOTTING INPUT IMPEDANCE

Input impedance is, quite often, an important circuit parameter. Often it will vary with frequency. Without going into the problem of designing required impedance levels, let's look at how PSpice may be used to simulate and display input impedance.

Impedance is similar to resistance in that it is a coefficient in the observed change in voltage and current, so impedance is a complex resistance. From Ohm's law we know that R = V/I. In a similar fashion, if V is a complex voltage

$$V = a + i \cdot b = |V| e^{i \angle V}$$

(where $|V|$ and $\angle V$ are the magnitude and phase angle of V) and I is a complex current

$$I = c + i \cdot d = |I| e^{i \angle I}$$

(where $|I|$ and $\angle I$ are the magnitude and phase angle of I), then impedance is

$$Z = V/I = (a + i \cdot b)/(c + i \cdot d) = ((a \cdot c + b \cdot d) + i \cdot (b \cdot c - a \cdot d))/I^2 = (|V|/|I|) e^{i(\angle V - \angle I)}$$

By using this result we may directly plot the magnitude, phase, real part, or imaginary part of input or output impedance. Using the same LC-filter circuit from before, we can plot input impedance without even rerunning the simulation. All of the data has already been calculated from the simulations for output response. The magnitude of the input impedance may be plotted using:

v(1)/i(r1) is $|Z_i|$ for $Q = \frac{1}{2}$
v(1)/i(r2) is $|Z_i|$ for $Q = 1$
v(1)/i(r3) is $|Z_i|$ for $Q = 2$
v(1)/i(r4) is $|Z_i|$ for $Q = 4$

See Fig. 6.5. Note that v(1) is the magnitude of the input voltage, which was set to a constant value which could have been used in the formula.
Note that the Y-axis was set to a logarithmic scale. This has the same effect as displaying, say, output response in decibels (as in a "Bode" plot). You can see the "peaking" of the input impedance for $Q > 1$, and the asymptotic trend of impedance as the frequency tends toward low or high frequencies.

The phase angle of the input impedance may be plotted using:

−ip(r1) is $\angle Z_i$ for $Q = \frac{1}{2}$
−ip(r2) is $\angle Z_i$ for $Q = 1$
−ip(r3) is $\angle Z_i$ for $Q = 2$
−ip(r4) is $\angle Z_i$ for $Q = 4$

See Fig. 6.6. In this case, we "cheat" by knowing that the phase angle of the input voltage is zero.

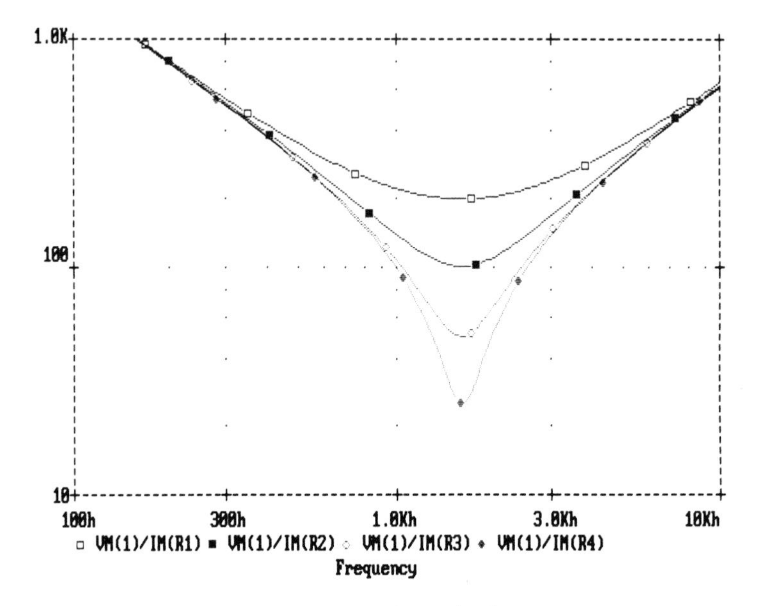


**Figure 6.5** Plot of |Zin|.

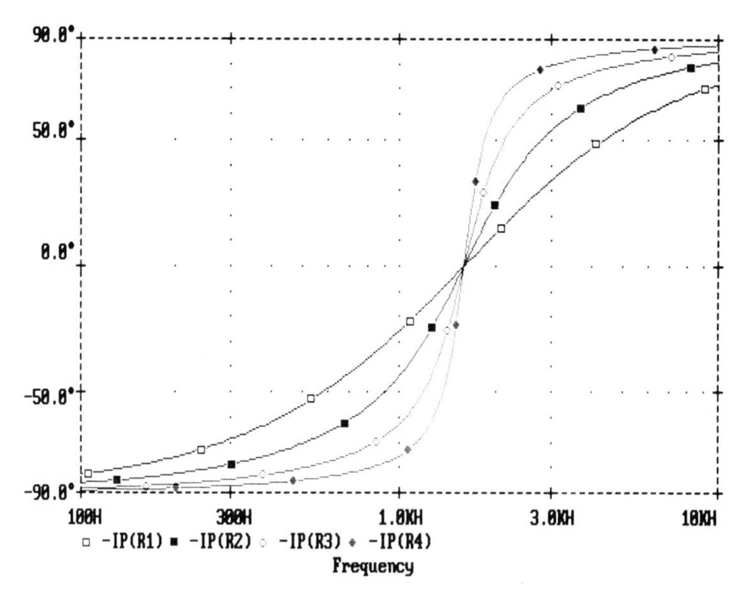


**Figure 6.6** Plot of ∠Zin.

**Exercise 6.7.1**

Develop a complete formula for the phase of an impedance. Use the real and imaginary parts of the voltage across the impedance and the current flowing through the impedance.

**Exercise 6.7.2**

Using the LC-filter circuit simulation, display Probe output for the following values: VI(1) and VI(2). What does this tell you about the input impedance? Now, after clearing the display, plot these values: I(R1), IR(R1), and II(R1). Where does the imaginary part of the input current cross the X-axis? What is significant about this frequency value? Try plotting the formula for the magnitude of the input current using the real and imaginary parts (only).

The real part of the input impedance may be plotted using:

$$v(1) \cdot ir(r1)/(i(r1) \cdot i(r1)) \text{ is } re(Z_i) \text{ for } Q = \tfrac{1}{2}$$

$$v(1) \cdot ir(r2)/(i(r2) \cdot i(r2)) \text{ is } re(Z_i) \text{ for } Q = 1$$

$$v(1) \cdot ir(r3)/(i(r3) \cdot i(r3)) \text{ is } re(Z_i) \text{ for } Q = 2$$

$$v(1) \cdot ir(r4)/(i(r4) \cdot i(r4)) \text{ is } re(Z_i) \text{ for } Q = 4$$

This time we "cheat" by knowing the input voltage source has no imaginary part from the AC specification, which is another way of saying that its phase angle is zero.

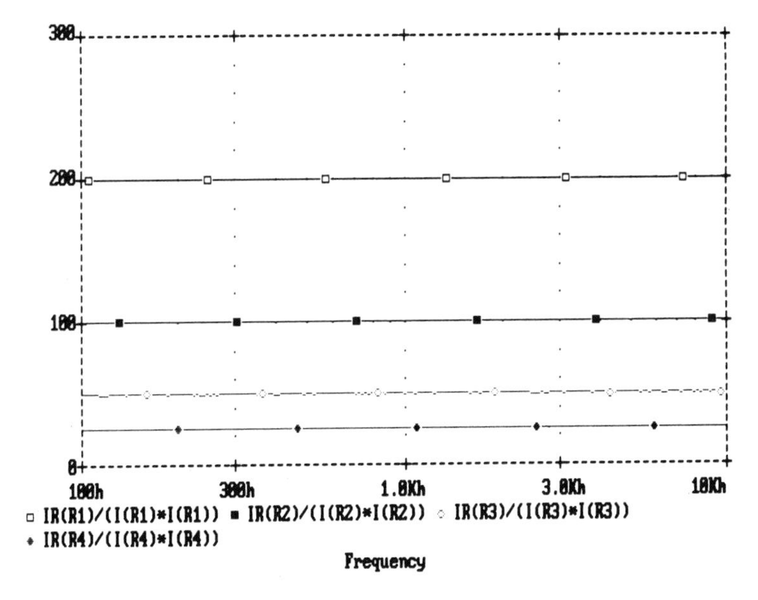


**Figure 6.7** Plot of *re*(Zin).

**Exercise 6.7.3**

Develop a complete formula for the real part of the input impedance. In this case, the magnitude and phase of the input voltage are unknown.

The imaginary part of the input impedance may be plotted using:

$$-ii(r1)/(i(r1)\cdot i(r1)) \text{ is } im(Z_i) \text{ for } Q = \tfrac{1}{2}$$

$$-ii(r2)/(i(r2)\cdot i(r2)) \text{ is } im(Z_i) \text{ for } Q = 1$$

$$-ii(r3)/(i(r3)\cdot i(r3)) \text{ is } im(Z_i) \text{ for } Q = 2$$

$$-ii(r4)/(i(r4)\cdot i(r4)) \text{ is } im(Z_i) \text{ for } Q = 4$$

Again, we "cheat" by knowing the input voltage source has no imaginary part. Furthermore, we also realize that the real part of the input voltage is unity.

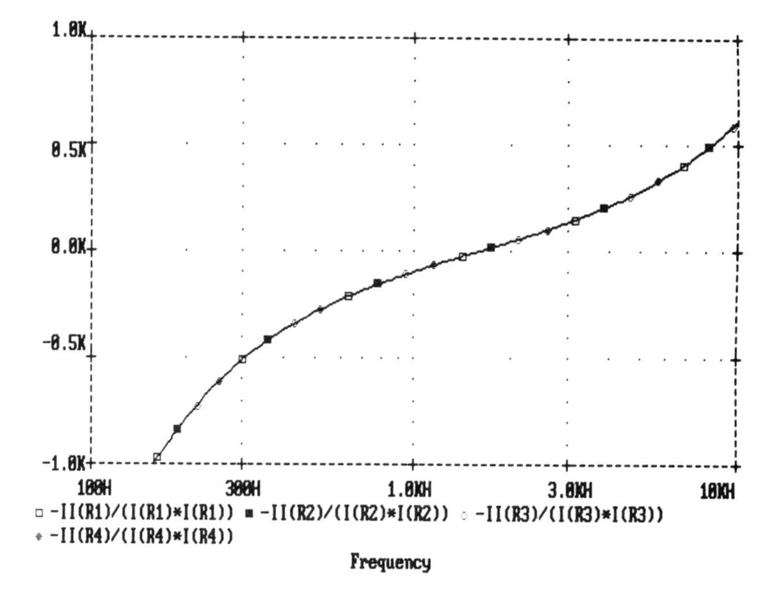


**Figure 6.8** Plot of *im*(Zin).

**Exercise 6.7.4**

Develop a complete formula for the imaginary part of the input impedance. As before, the magnitude and phase of the input voltage are unknown.

**Exercise 6.7.5**

You may have noticed that all of the plots for the imaginary part of the input impedance were the same. This means that the complex impedance part of the circuit was unchanged, regardless of which Q values we were investigating. Reimplement the Q = 4 circuits with real input impedance of 50 ohms and the same resonant frequency as the example. Then, plot the imaginary parts of the input impedance. Compare it to the example for Q = 2.

## 6.8 PLOTTING OUTPUT IMPEDANCE

Output impedance is, quite often, another important circuit parameter. We can use PSpice to simulate voltages and currents and then use Probe to display formulations for output impedance, just as we did previously for input impedance. You can even use these techniques for deriving the impedance levels internal to circuits, so long as you are careful to include the correct voltages and currents.

We were lucky, in the previous example, that one simulation run provided everything we needed for output response and a variety of ways of looking at input impedance. Applying a voltage source to the input was necessary to excite the circuit for the output response simulation, and accordingly supplied current to the circuit from which the impedance calculations were done. To look at output impedance we need to have current flowing at the output terminals to calculate impedance levels.

For the LC-filter circuit, the input impedance calculations assumed that the output was not connected to a load. This is the same as saying the load had "infinite" impedance. The output impedance of the circuit will assume that the input is connected to a voltage source, which is to say that the input will have a load of zero impedance (the input is, for frequency calculations, "shorted"). To make the appropriate changes to our circuit file we will need to set the input voltage to zero

```
Vin 1 0 0
```

and apply some current generators at the output of each filter section

```
I1 0 3 AC 1
I2 0 5 AC 1
I3 0 7 AC 1
I4 0 9 AC 1
```

Note that the current sources are connected so that positive current flows into the output node of the LC-filter. This was done so the phasing of the current will be intuitively correct when we display the output impedance. Now, we rerun the simulation.

The formulas we developed for input impedance are the same for output impedance, although the observations about which components of the voltage source are zero or unity now apply to the current source. The magnitude of the output impedance may be plotted using:

$$v(3)/i(i1) \text{ is } |Z_0| \text{ for } Q = \tfrac{1}{2}$$

$$v(5)/i(i2) \text{ is } |Z_0| \text{ for } Q = 1$$

$$v(7)/i(i3) \text{ is } |Z_0| \text{ for } Q = 2$$

$$v(9)/i(i4) \text{ is } |Z_0| \text{ for } Q = 4$$

Note that i(i1), i(i2), i(i3), and i(i4) are the magnitude of the driving current, which was set to a constant value that could have been used in the formula.

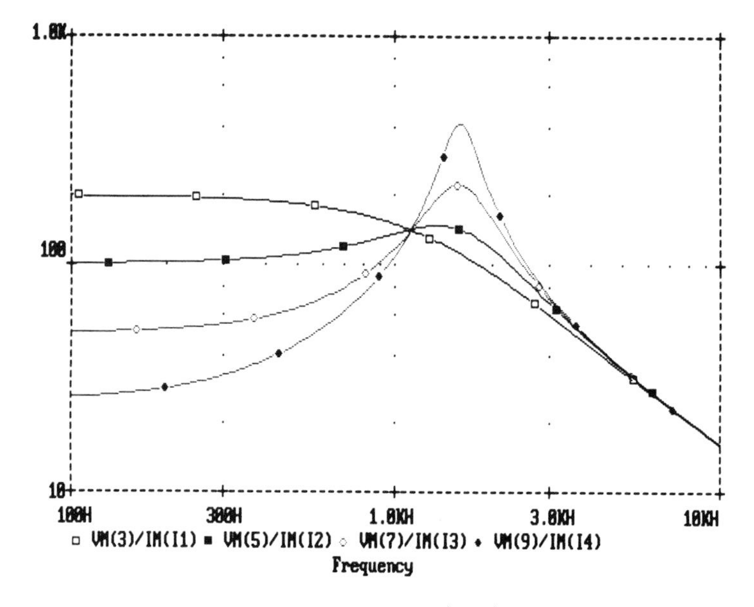


**Figure 6.9** Plot of |Zout|.

Note that the Y-axis was set to a logarithmic scale. This has the same effect as displaying, say, output response in decibels. You can see the "peaking" of the output impedance for $Q > 1$, and the asymptotic trend of impedance as the frequency tends toward low or high frequencies.

The phase angle of the output impedance may be plotted using:

$$vp(3) \text{ is } \angle Z_0 \text{ for } Q = \tfrac{1}{2}$$

$$vp(5) \text{ is } \angle Z_0 \text{ for } Q = 1$$

$$vp(7) \text{ is } \angle Z_0 \text{ for } Q = 2$$

$$vp(9) \text{ is } \angle Z_0 \text{ for } Q = 4$$

In this case, we "cheat" by knowing that the phase angle of the output current is zero (from the AC statement).

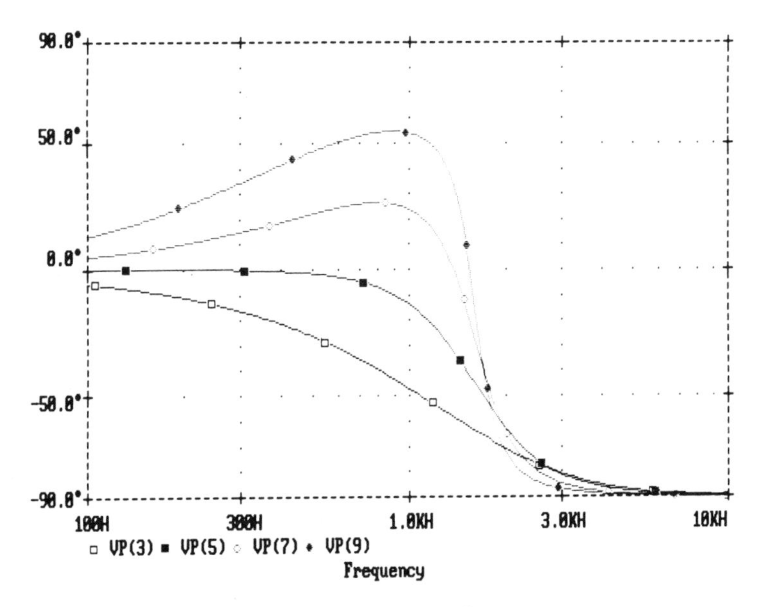


**Figure 6.10** Plot of ∠Zout.

The real part of the output impedance may be plotted using:

vr(3)/(i(i1)·i(i1)) is $re(Z_0)$ for $Q = \frac{1}{2}$

vr(5)/(i(i2)·i(i2)) is $re(Z_0)$ for $Q = 1$

vr(7)/(i(i3)·i(i3)) is $re(Z_0)$ for $Q = 2$

vr(9)/(i(i4)·i(i4)) is $re(Z_0)$ for $Q = 4$

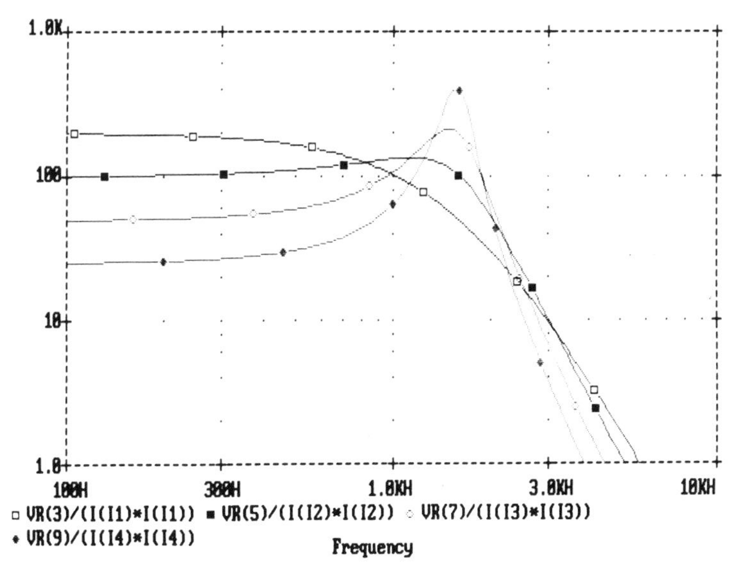


**Figure 6.11** Plot of *re*(Zout).

Again we "cheat" by knowing the driving current source has no imaginary part, which is another way of saying that its phase angle is zero.

**Exercise 6.8.1**

Develop a complete formula for the real part of the output impedance. In this case, the magnitude and phase of the output voltage are unknown.

The imaginary part of the output impedance may be plotted using:

vi(3)/(i(i1)·i(i1)) is $im(Z_0)$ for $Q = \frac{1}{2}$

vi(5)/(i(i2)·i(i2)) is $im(Z_0)$ for $Q = 1$

vi(7)/(i(i3)·i(i3)) is $im(Z_0)$ for $Q = 2$

vi(9)/(i(i4)·i(i4)) is $im(Z_0)$ for $Q = 4$

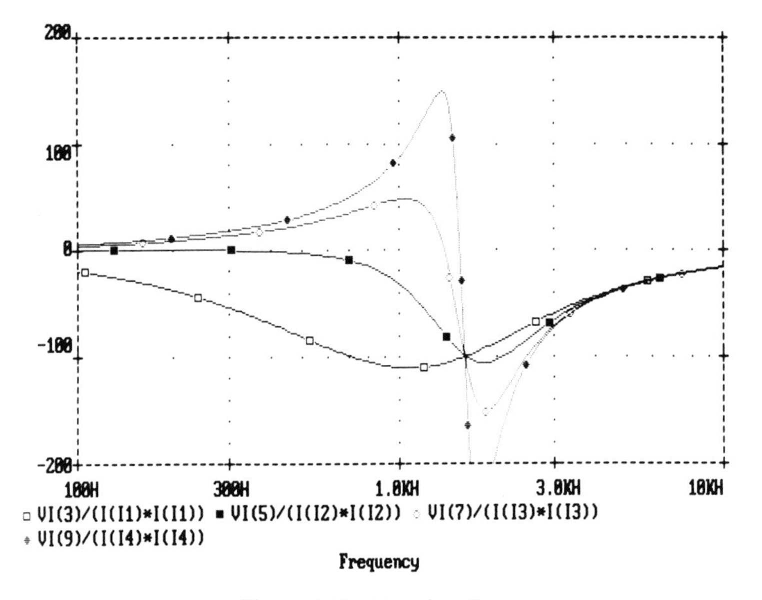


**Figure 6.12** Plot of *im*(Zout).

Again, we "cheat" by knowing the driving current source has no imaginary part. Also, we take advantage of the real part of the driving current being unity.

**Exercise 6.8.2**

Develop a complete formula for the imaginary part of the output impedance. As before, the magnitude and phase of the output voltage are unknown.

## 6.9 PLOTTING LOOP GAIN

Inspecting the open-loop gain of circuits with high-gain components is a difficult task. You might be tempted to open the loop to make measurements, but this will probably destroy the DC bias of the circuit. Opening the loop might also disconnect an internal load impedance and affect your measurements. What we need is a way to make these measurements without opening the loop or changing the internal loading of the circuit.

In developing this technique we will start with a hypothetical circuit. This circuit will be set up to make these measurements without introducing any problems. Then we will modify the formulas to work with "real" circuits. But first, let us review some terms.

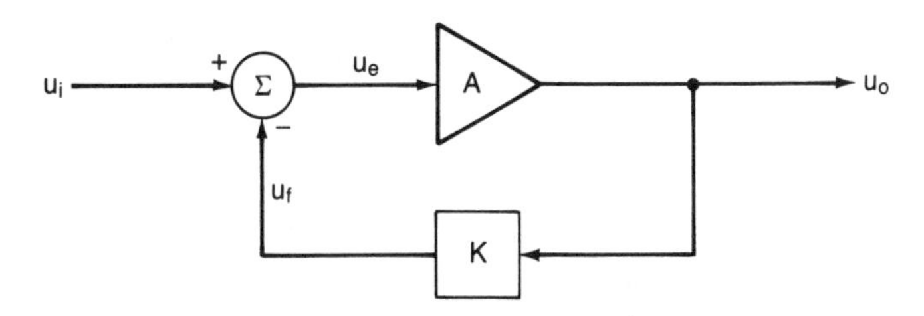


**Figure 6.13** Schematic of system with feedback.

In the diagram of the system with feedback, the following signals may be calculated by inspection

$$u_0 = A \cdot u_e$$

$$u_e = u_i - K \cdot u_0$$

where $u_i$, $u_0$, and $u_e$ are the "input," "output," and "error" signals, respectively. Further manipulation of these formulas yields

$$u_0 = A \cdot u_i - A \cdot K \cdot u_o$$

$$u_e = u_i/(1 + A \cdot K)$$

so that system gain is

$$G = u_0/u_i = A/(1 + A \cdot K) = A/(1 + T) = (1/K) \cdot T/(1 + T)$$

and the "open-loop gain," or "return ratio," is

$$T = u_f/u_e = A \cdot K$$

The system's relative stability can be inspected in an open-loop configuration. The loop is opened in the feedback path, and a "test" signal is injected. Then the resulting feedback signal, opposite the injection point, is compared to the test signal. The feedback signal is inspected for one full cycle of phase shift (actually only

180° as the feedback is subtracted). If the open-loop gain (the ratio of feedback signal to the test signal) is one, or greater, then the loop is unstable since it can supply its own input. The amount of gain, relative to unity, at 180° phase shift is called the open-loop "gain margin." Likewise, the amount of phase difference from 180°, when the open-loop gain is unity, is called the open-loop "phase margin."

Actually this analysis is true even if the loop is broken in the forward path, but it seems easier to describe as though the feedback path were broken. Now, having said that, we will imagine that we have "broken" the loop somewhere inside the circuit. It doesn't matter where, just as long as the break is in some part of the loop. To do this, we will imagine that some part of the signal path is a controlled-current source connected to an impedance (for example, this may be a transistor with a load resistance) as shown in Figure 6.14(a).

Then we break the loop, as shown in Figure 6.14(b), being careful to duplicate the load impedance seen by the current source. Now we inject a current into the original load impedance, which is part of the original loop, as the test signal. The return signal is current in the duplicated load impedance. The ratio of the two signals is called "open-loop current gain," or Ti, where

$$Ti = i_y/i_x$$

Now comes part of the trick. It is not necessary to open the loop to inject the test signal. As shown in Figure 6.14(c), if a current is injected into the signal path it splits into the test signal and return signal of Figure 6.14(b). Furthermore, the load impedance does not need to be duplicated, since the loop is not broken and the current source has infinite impedance.

Similarly, we can develop the same technique using voltages instead of currents.

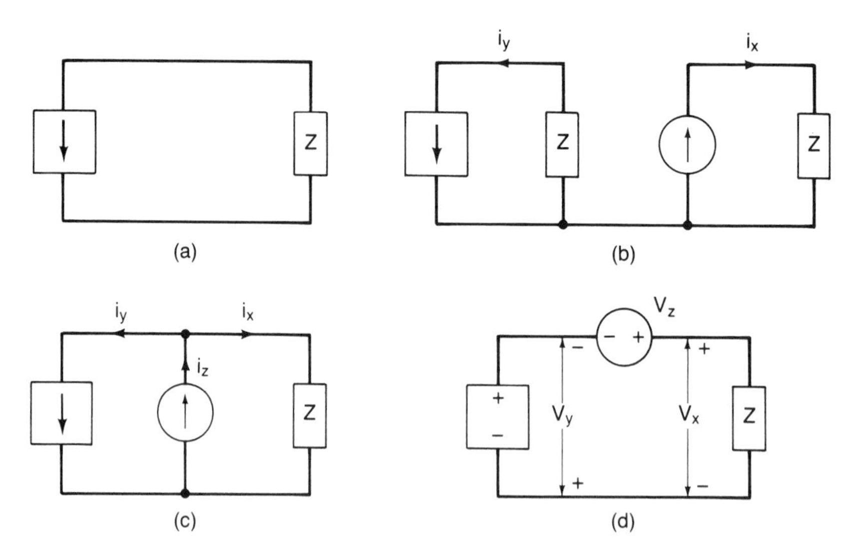


**Figure 6.14** Schematics showing development of open-loop measurements.

As shown in Figure 6.14(d), a voltage source is inserted in the loop to inject a signal. The resulting measurement of the voltage across the load impedance and the controlled-voltage source yields a ratio called the "open-loop voltage gain," or Tv, where

$$Tv = v_y/v_x$$

Since we were able to choose ideal points to break the loop, these measured loop gains are the open-loop gain of the system, that is T = Ti for the current-source version, and T = Tv for the voltage-source version.

Now we tackle a more "real" circuit, as our example so far assumed controlled sources that had no internal impedance the way a real transistor, or real opamp, does. We can account for this by use of superposition.

In the previous case the injection was done as shown in Figure 6.15(a), but in this case we have two new currents, as shown in Figure 6.15(b), and

$$i'_y = M \cdot i'_x + i'_x \cdot z_1/z_2$$

where M is the rest of the loop's gain. The open-loop current ratio that we measure is, then

$$i'_y/i'_x = Ti = M + z_1/z_2$$

and the open-loop gain is

$$T = i_y/i_x$$
$$T = M \cdot i'_x/(i'_x \cdot z_1/z_2 + i'_x)$$
$$T = (M + z_1/z_2 - z_1/z_2)/(1 + z_1/z_2)$$
$$T = (Ti - z_1/z_2)/(1 + z_1/z_2)$$

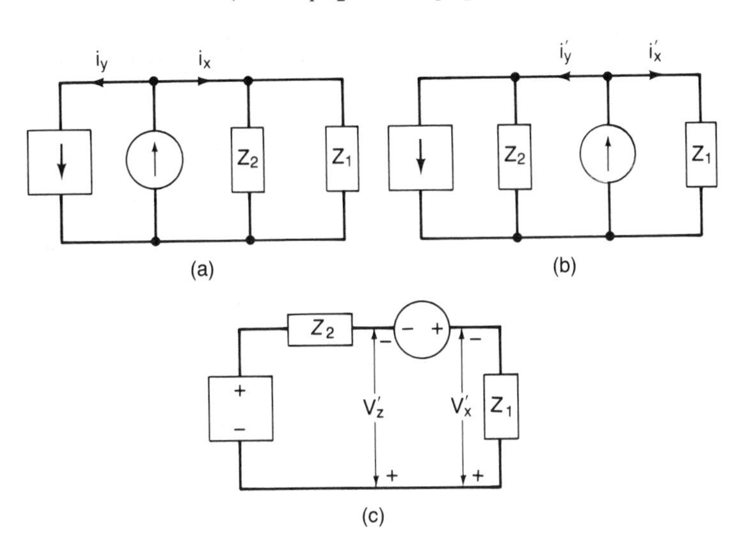


**Figure 6.15** Schematics showing open-loop measurements with internal impedances.

By similar means, we might make a non-ideal measurement of the open-loop voltage ratio, as shown in Figure 6.15(c), with the result

$$T = (Tv - z_2/z_1)/(1 + z_2/z_1)$$

Of course, this begs the question of what the circuit impedances are so we may calculate T. By inspection, we can see that

$$Ti \approx T \text{ if } z_1/z_2 << 1$$

and

$$Tv \approx T \text{ if } z_2/z_1 << 1$$

but, if we were to measure both Ti and Tv, then we would have two equations with two unknowns. Eliminating the ratio, $z_1/z_2$, yields

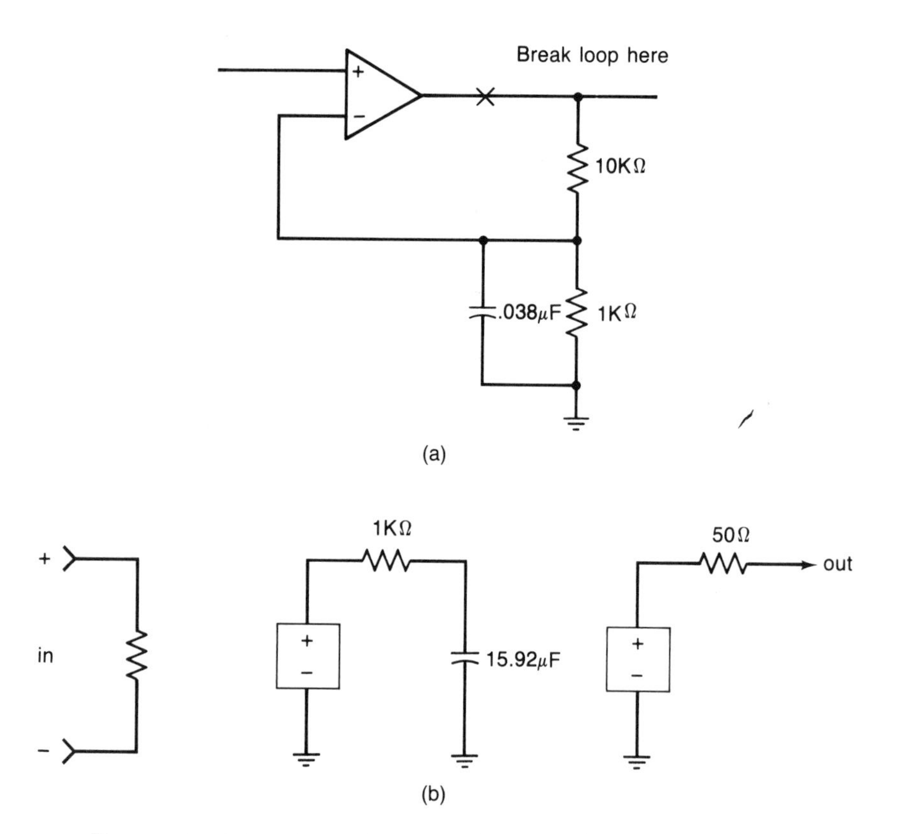


**Figure 6.16** Schematics for circuit example of open-loop gain measurements.

$$(T + 1) = (Ti + 1) || (Tv + 1)$$

where || means "parallel combination," for example

$$x || y = 1/(1/x + 1/y)$$

This says that the lower of the two measurements, Ti or Tv, dominates the value of T, the open-loop gain. Another way to restate the formula for T is

$$T = (Ti \cdot Tv - 1)/(Ti + Tv + 2)$$

Let's try to measure open-loop gain in a relatively simple circuit. Figure 6.16(a) shows the circuit we will measure, breaking into the loop at the output of the opamp.

The simplified model of the opamp, shown in Figure 6.16(b), will be used for this example. The subcircuit definition for the opamp is

```
* "ideal" op-amp with 100K gain and one-pole roll-off at 10Hz
.subckt opamp non inv out
  rin non inv 100K
  egain 1 0 (non, inv) 100K
  ropen 1 2 1K
  copen 2 0 15.92u
  eout 3 0 (2,0) 1
  rout 3 out 50
.ends
```

Since we will want two copies of the entire circuit we are measuring, let's put the circuit in a subcircuit. This subcircuit will have only two nodes, which are at the place where we are breaking the loop.

```
* example circuit
.subckt test in out
  vin 1 0 0 ; zero input
  x1 1 2 in opamp
  r1 out 2 10K
  r2 2 0 1K
  c1 2 0 .038u
.ends
```

Finally, there is the rest of the circuit.

```
Open-loop gain measurement

x1 Tii Tio test ; this copy for Ti measurements
X2 Tvi Tvo test ; this copy for Tv measurements

* perform Ti measurements
iz 0 1 AC 1 ; "test" current generator
v_ix 1 Tio 0 ; sense Ix
v_iy 1 Tii 0 ; sense Iy
h_ix ix 0 v_ix 1 ; convert Ix to a voltage
r_ix ix 0 1G
h_iy iy 0 v_iy 1 ; convert Iy to a voltage
r_iy iy 0 1G

* perform Tv measurements
vz Tvo Tvi AC 1 ; "test" voltage generator
e_vx vx 0 Tvo 0 1 ; copy Vx
r_vx vx 0 1G
e_vy vy 0 0 Tvi 1 ; copy Vy
r_vy vy 0 1G
```

After running an AC analysis simulation, we view the results.

First we look at the magnitude of Ti and Tv. As we would expect, Tv dominates—that is, Tv is of smaller magnitude and will control the value of (Ti + 1)||(Tv + 1)—since the input impedance of the feedback circuit is much greater than

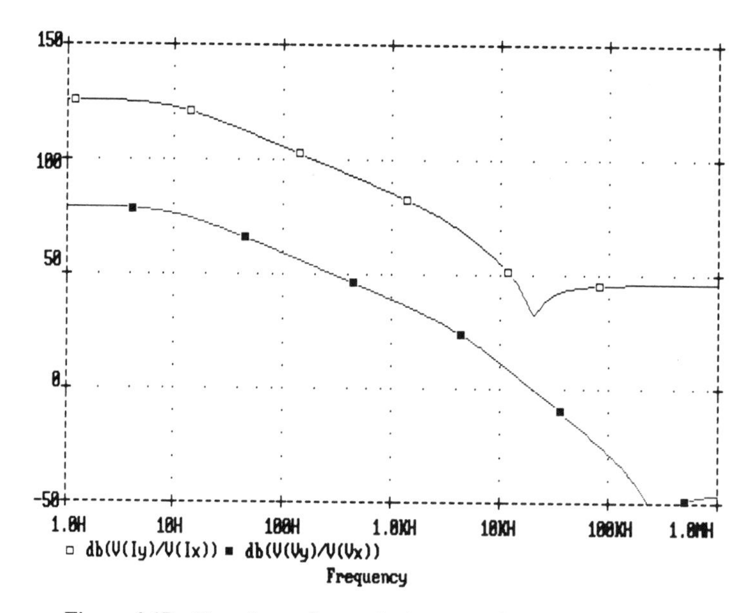


**Figure 6.17** Plot of open-loop gain (current gain and voltage gain).

the output impedance of the opamp. From this we gain the insight that even though we may be able to calculate T from Ti and Tv, it is better to have Ti or Tv dominate so that no calculation is necessary. This means that we should try to break the loop where the return side is a relatively ideal current, or voltage, source.

Looking at Tv more closely, we can put up another plot of the phase of Tv. The phase margin for the system will be the phase at unity loop-gain, plus 180°.

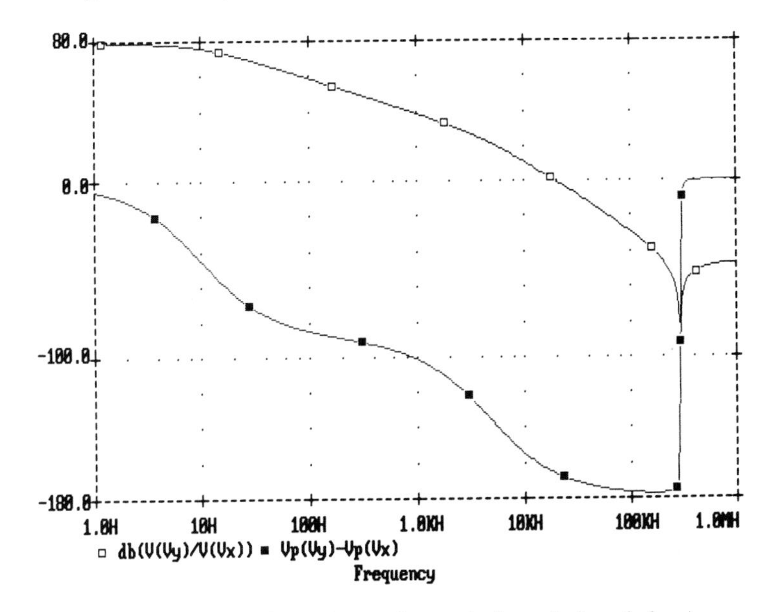


**Figure 6.18** Plot of open-loop voltage gain (magnitude and phase).

Notice that we may break into the loop of this example circuit elsewhere, or try another circuit, just by changing the description of the test subcircuit.

# CHAPTER 7

# Feedback Control Analysis

As it happens, the first applications of feedback control theory were in electrical systems, in contrast to earlier mechanical systems which were often designed using Routh's method as the only analytical tool. During the early days of electrical and electronic systems a great amount of effort was spent on analytical techniques for these systems. The result was a variety of graphical tools for analyzing the frequency response of systems and synthesizing stable control systems, both electrical and mechanical. The remarkable outcome of these tools is the quick, graphic analysis of transient (time-domain) response and stability from frequency response. Moreover, that closed-loop frequency response can be deduced from open-loop frequency response.

Most physical systems, or "plants" (a control-theory idiom), may be analyzed by PSpice by creating an electrical analog for the controlling equations: force translates into current, mass translates into capacitance, and so forth. Then the frequency response of the system (linear displacement, rotation, etc.) may be calculated using PSpice's frequency response analysis. As mentioned in the introduction to this book, PSPice is not concerned that your electrical analog is not even a normal electronic circuit. By using this technique of "analysis by analogy" you may even directly convert system equations in "*s*" to electrical equivalents for PSpice.

## 7.1 DYNAMIC PLANT EXAMPLE

Let us consider a general example of a linear system with feedback control, as shown in Figure 7.1.

This system has a "dynamic plant" and simple feedback; it has (i) an input

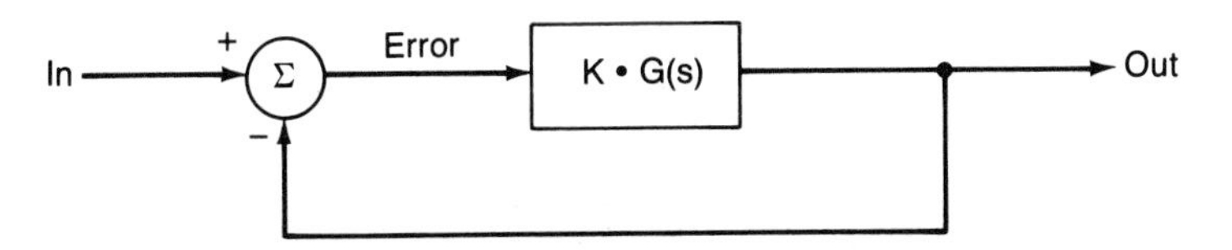


**Figure 7.1** Schematic of generalized feedback system.

signal, X, (ii) an error signal, E, (iii) a "plant" function, K · G(s), which separates the linear gain from the complex part of the plant, and (iv) an output signal, Y.

Of course, when the loop is open the error signal is the same as the input signal. Functions of these signals are calculated to create the various plots in this chapter; for example:

$$|Y|/|X| = \text{system response magnitude}$$

$$\underline{/Y} - \underline{/X} = \text{system phase response}$$

Let us look at an example circuit and see what feedback system analysis plots result from its frequency response.

Figure 7.2 shows a more complicated example of the two-pole filter we analyzed in Chapter 6 on frequency response. The circuit file that describes this circuit is

```
2 tandem, double-pole, low-pass, LC-filters
Vin 1 0 AC 1
Rin 1 0 1K
Ein in 0 poly(2) (1,0) (6,0) 0 2 0
*Q=4
R1 in 2 25
L1 2 3 10mH
C1 3 0 1uF
*Q=4 @ 1/5 frequency
E2 4 0 3 0 1
R2 4 5 25
L2 5 6 50mH
C2 6 0 5uF
.AC DEC 100 100hz 10Khz
.PROBE
.END
```

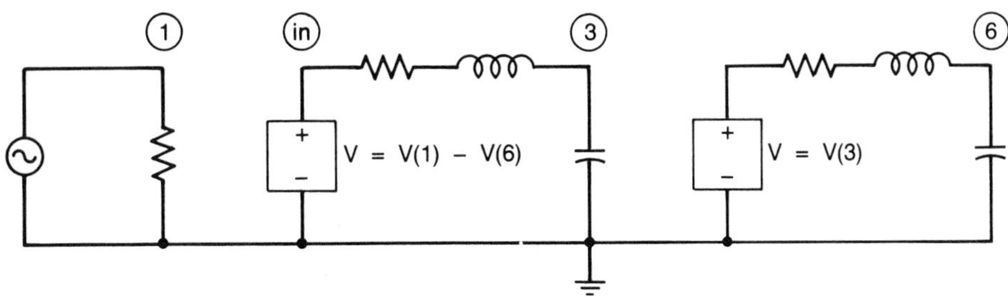


**Figure 7.2** Schematic of tandem LC circuit with gain.

You will notice that we have placed, in series, two of the Q = 4 RLC filter sections we analyzed earlier. We have also used an E-device to serve as an ideal difference amplifier, which lets us change the gain easily and choose to close the feedback loop (that is, if the gain from the feedback input is zero then the circuit is operating "open-loop"). This circuit does not correspond to any useful circuit or physical system analogy. However, we can still look at the system response.

## 7.2 BODE PLOTS

An important theorem by H. W. Bode deals with linear systems with constant coefficients whose poles and zeros are all in the left-half of the *s*-plane (minimum-phase systems). This theorem holds that, for any minimum-phase system, the phase-angle part of the frequency response is uniquely related to the magnitude part. This allows the phase response of a circuit to be deduced from the magnitude response, and vice versa. Moreover, entire systems of coupled minimum-phase sections could be analyzed by a graphical technique that embodies Bode's theorem. This technique was extremely useful in electronic circuits, where complex systems can be easily built but are potentially more difficult to analyze. The graphical technique used what came to be called "Bode plots."

Bode plots are graphs of the magnitude and phase response of the circuit versus frequency of sinusoidal excitation (that is, small-signal analysis, or what SPICE calls frequency response). We have seen these plots in the previous chapter. The simplicity of construction for these plots, **when done by hand,** is due to the

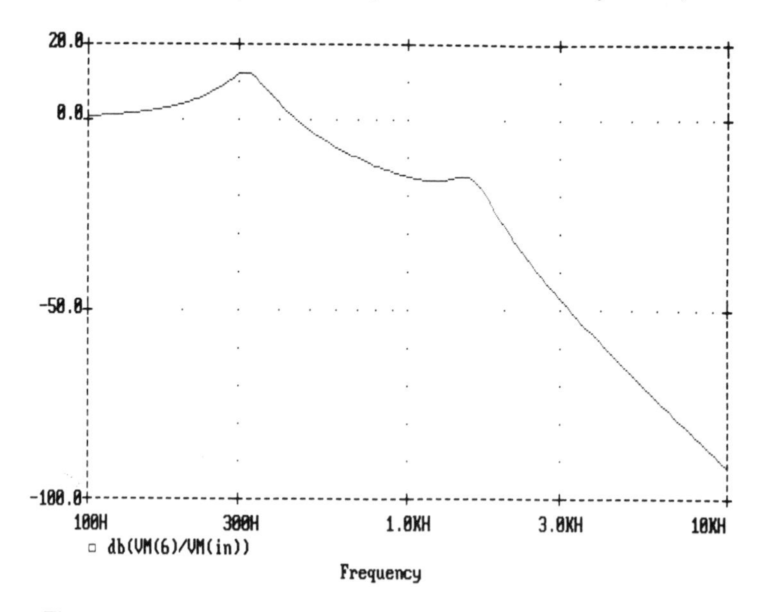


**Figure 7.3** Plot of system's open-loop magnitude response (in decibels).

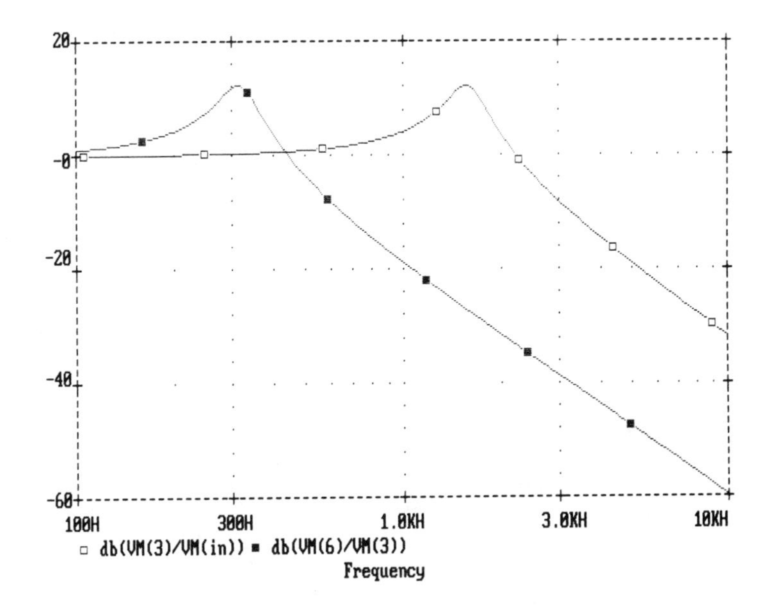


**Figure 7.4** Plot of the individual section's magnitude response (in decibels).

frequency scale (which is along the X-axis) being logarithmic. Phase angle is plotted linearly along the Y-axis, but magnitude is plotted with a logarithmic scale along the Y-axis. This setup allows ratios of magnitude and ratios of frequency to have constant displacement on the graph.

The ease in constructing Bode plots made them popular and standard. PSpice will create them for you either by using the .PLOT statement or, more usefully, by using Probe graphics. Now, after running PSpice on this circuit we may run Probe to look at some Bode plots.

See Figure 7.3. This plot shows the open-loop magnitude response of the system. We can see that system response is actually the product of its components: a gain section, a lower-frequency two-pole filter, and a higher-frequency two-pole filter. We may plot these responses separately, as shown in Figure 7.4.

Phase plots are also simple, as shown in Figure 7.5.

The total phase is the sum of the phase of the individual sections, as shown in Figure 7.6.

These examples show what made the Bode plot technique so useful and popular: the ease with which graphs of the response of each section's circuitry could be constructed and then combined to produce the system response graph. Of course, this was all being done **by hand,** but now you can use PSpice.

One of the items not handled well by Bode plots was deducing closed-loop response from open-loop response. We can resimulate our circuit in the closed-loop configuration by modifying the statement for the difference amplifier (E-device) as follows:

```
Ein in 0 poly(2) (1,0) (6,0) 0 2 0
```

becomes

```
Ein in 0 poly(2) (1,0) (6,0) 0 2 -2
```

to close the loop. After running PSpice we may compare the open-loop and closed-

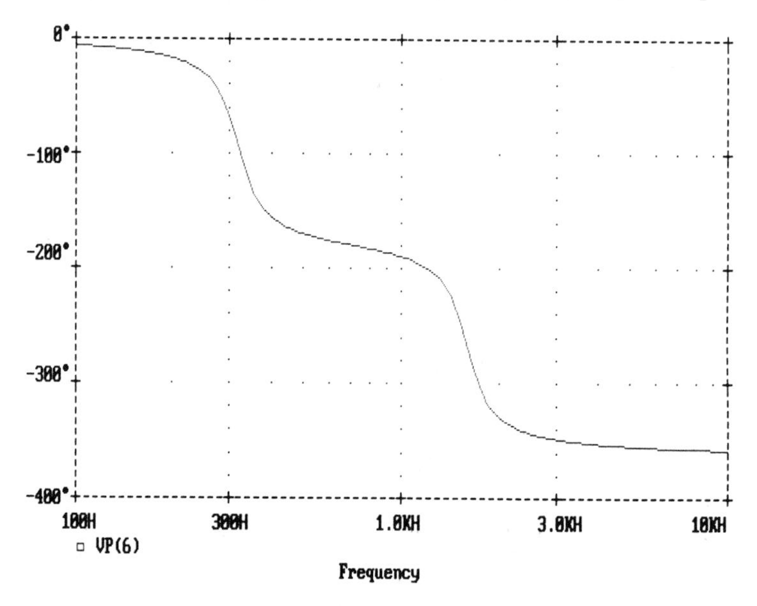


**Figure 7.5** Plot of the system's open-loop phase response.

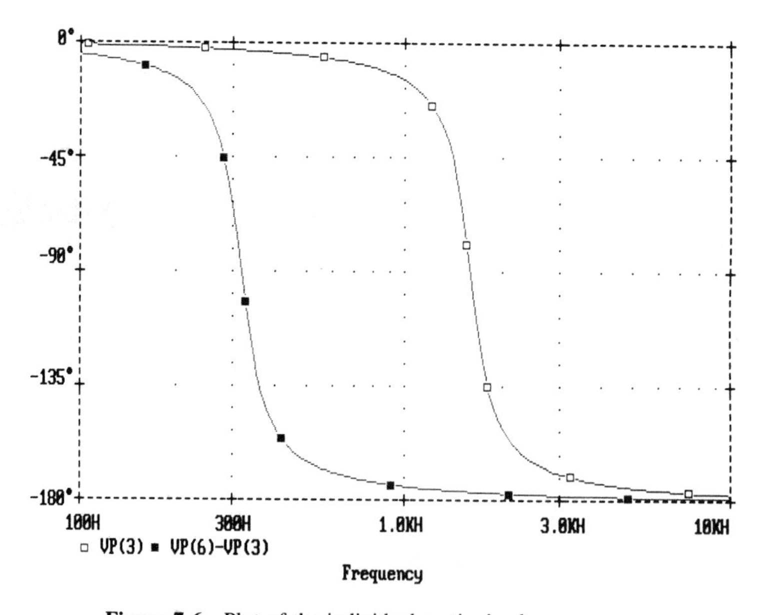


**Figure 7.6** Plot of the individual section's phase response.

loop responses. You will notice that the system gain peaks at a new frequency and is the frequency at which the open-loop response crossed Odb!

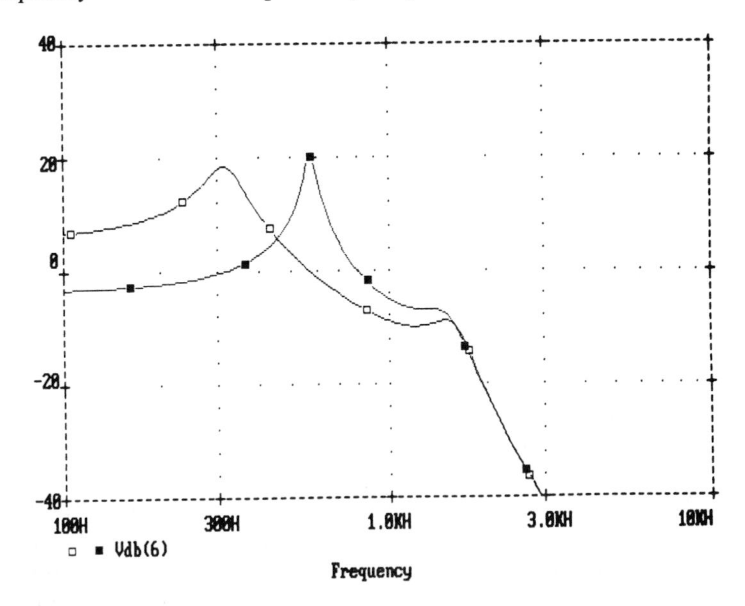


**Figure 7.7** Plot of the system's open-loop and closed-loop magnitude response (in decibels).

As you might expect, the phase response of the system has changed as well.

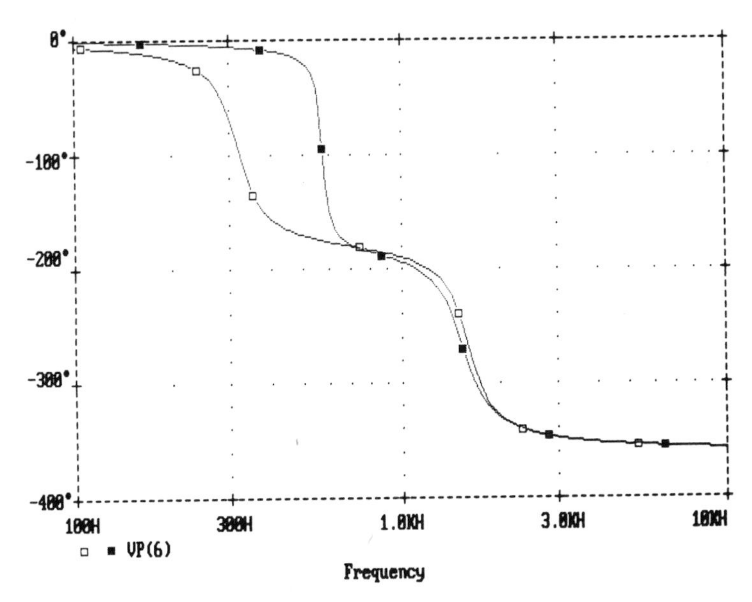


**Figure 7.8** Plot of the system's open-loop and closed-loop phase response.

In the next section we will look at another type of plot that does deduce closed-loop response, from the open-loop response, by merely shifting the origin of the plot.

## 7.3 INVERSE-POLAR PLOTS

Another graphical technique, developed by H. Nyquist, uses a polar coordinate graph to plot system response characteristics. From the mathematics of feedback systems, we know that the closed-loop system response is

$$Y/X = K \cdot G(s)/(1 + K \cdot G(s))$$

The magnitude and phase of this response can be plotted for various frequencies, as with the Bode plot, if we could plot this using polar coordinates. In this case the magnitude of the response would be the magnitude of the vector, and the phase response would be the angle of the vector. Such a plot is quite easy to create with Probe, since the magnitude and phase of a vector is a function of the real and imaginary parts of the vector. This is called a ''polar plot'' and is created directly using the VR() and VI() display functions, which we used previously (for instance, when calculating impedances). By changing the X-axis of the plot to be the real component of the vector, and then plotting the imaginary component, the polar plot of the vector is created by using the rectangular components of the vector.

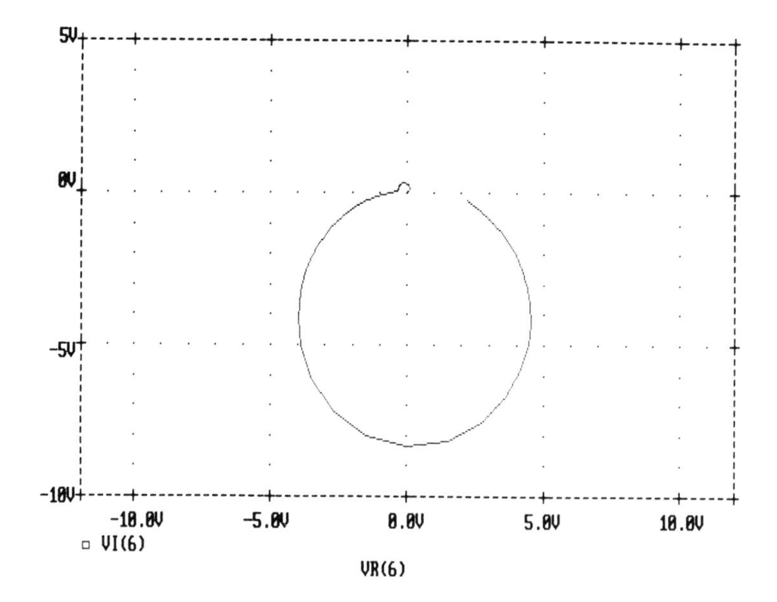


**Figure 7.9** Plot of system's closed-loop response in polar form.

The only parts missing from the plot are the markers for frequency.

In the feedback system equations, notice that

$$K \cdot G(s) = Y/E$$

and that the inverse of the system function, Y/X, is

$$X/Y = 1/(K \cdot G(s)) + 1$$

or

$$X/Y = E/Y + 1$$

which is very convenient, since E = X when the system is "open-loop." This means that the closed-loop response is related to the open-loop response by adding one, when the response is plotted as in inverse function. This is the utility of the "inverse polar" plot: the plot of the inverse transfer function is the same for both open- and closed-loop configurations; only the origin of the plot shifts. Let us try this with our example circuit.

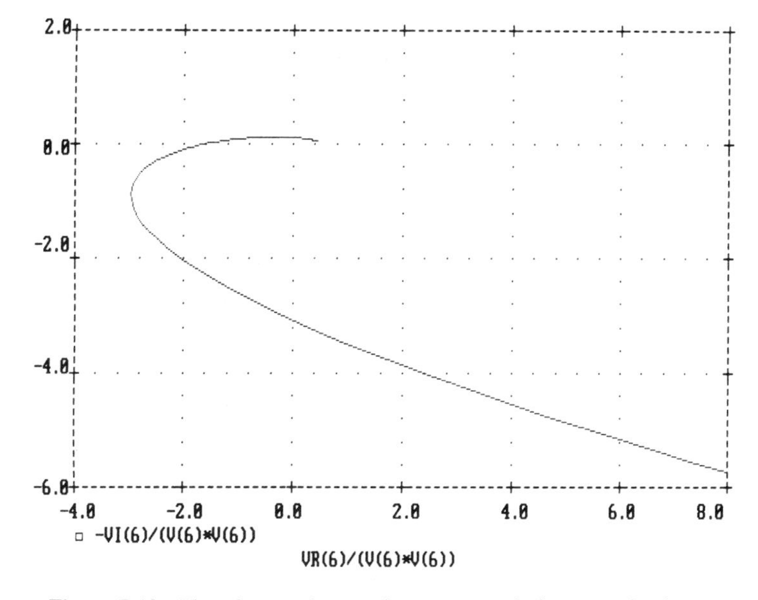


**Figure 7.10** Plot of system's open-loop response in inverse polar form.

Using the open-loop version of our circuit file (no feedback in Ein), our input is unity so the inverse of the output is (using complex arithmetic)

$$1/A = 1/(a + b \cdot i) = a/|A|^2 - b \cdot i/|A|^2$$

or, the coordinates for 1/Y are

$$re(Y)/|Y|^2, \ -im(Y)/|Y|^2$$

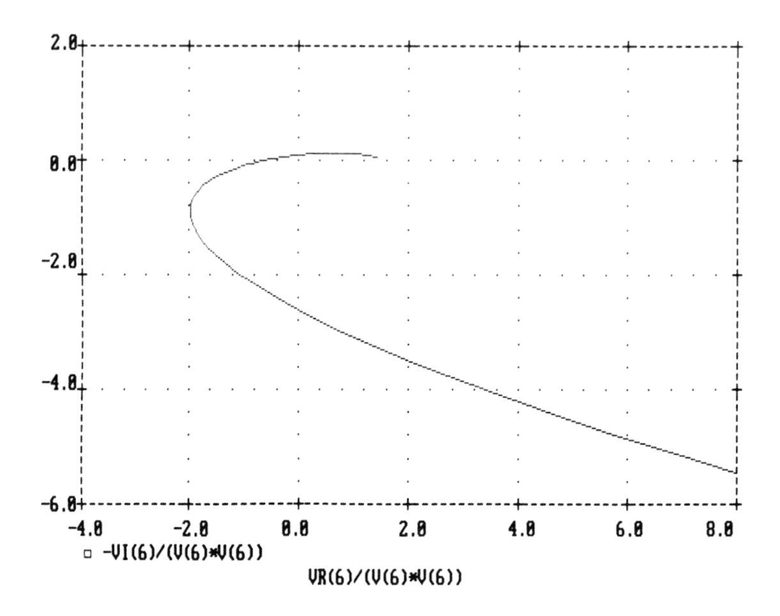


**Figure 7.11** Plot of system's closed-loop response in inverse polar form.

Now, we run PSpice again using the closed-loop version of the circuit file. Using the same display functions as before, the closed-loop response is plotted in Probe as shown in Figure 7.11.

If you are trying this, you will probably need to adjust the X-axis and Y-axis ranges, zooming in to the origin, to scale the plots for use and comparison. With everything adjusted, we notice that the curve is the same and only the origin of the plot has changed between the open-loop and closed-loop response plots.

**Exercise 7.3.1**

Check our equations and PSpice: directly evaluate E/Y + 1 in Probe, using complex arithmetic, and compare the open-loop response to the closed-loop response.

## 7.4 NICHOLS PLOTS

The Bode plot has the advantage of rapid plotting **by hand** using asymptote approximations and multiplying functions together by adding distances on a log scale. The inverse polar plot is more difficult to assemble by hand, but once plotted shows both the open-loop and closed-loop response. There is another graphical technique, developed by N. B. Nichols, which has both features.

The Nichols plot shows both magnitude and phase on the same plot, and is like a combined Bode plot (in fact, the easiest way to construct, by hand, a Nichols plot is to first construct a Bode plot and transcribe points to the Nichols plot).

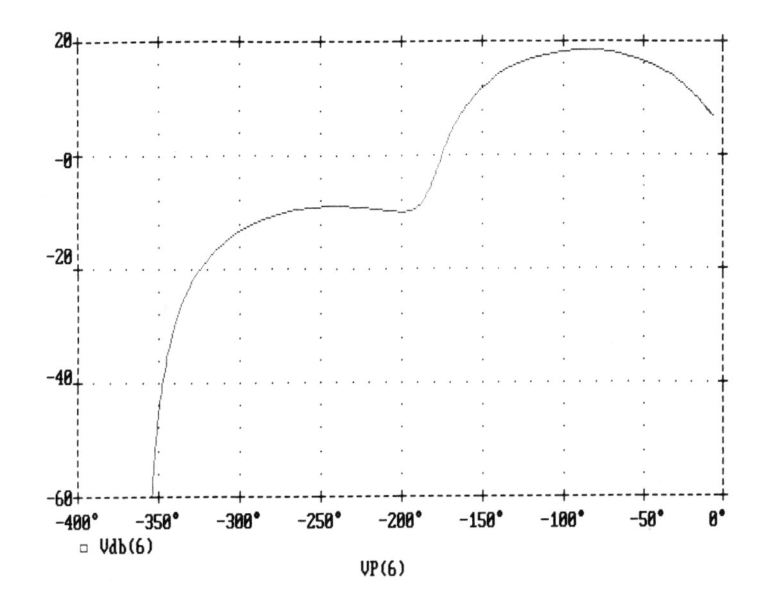


**Figure 7.12** Plot of system's open-loop response in Nichols form.

Using our example, we obtain the Nichols plot shown in Figure 7.12 by resetting the X-axis variable to be the phase response of the system.

The Nichols plot technique then uses an overlay plot to transform the open-loop response into a closed-loop response. This overlay plot is the same (so long as the plot coordinates are the same) for all system responses and performs the same job as shifting the origin for the inverse-polar plot.

# CHAPTER 8

# Noise Analysis

Whenever small signals are amplified or measured, you usually reach a lower limit of signal that is discernible; this limit is set by spontaneous fluctuations in the equipment you are using. The spontaneous fluctuations are called "noise," since, if the audio-frequency component of the fluctuating voltage, or current, were amplified and fed into a loudspeaker, you would hear a hissing noise. This type of fluctuation extends across all frequencies; for example, in television equipment noise creates the "snowy" picture.

PSpice can analyze noise by calculating the noise contributions from each element in the circuit and combining these noise sources with the various transfer functions in the circuit. You may print or plot (or use Probe) the result of these calculations to obtain noise response over a range of frequencies, just like the frequency response analysis.

## 8.1 NOISE CALCULATIONS

In PSpice, the resistors and semiconductor devices contribute to the noise calculations. While the semiconductor device noise models are more complicated, we can understand the general idea of noise analysis just by using resistors. In due course the semiconductor noise models will be explained.

The resistor, in PSpice, generates an equivalent "thermal noise" current in parallel with the resistor (which is then noiseless). Thermal noise is "white": its fluctuations contain equal amounts of all frequency components (which, in human vision, is white light). Technically, we would say thermal noise has "constant spectral density.'' The random fluctuations are characterized on a statistical basis using averages so that while the time average of a random fluctuation is zero, its mean-square average (or variance) has a value. The level of the resistor mean-square current generator is

$$\bar{i}^2 = 4 \cdot k \cdot T \cdot B/R$$

where $k$ = Boltzmann's constant: 1.38E−23 (W · sec/°K)

T = temperature in °K

B = bandwidth in hertz

R = resistance in ohms

Alternatively, the equivalent thermal noise could be represented by a mean-square voltage source in series with the resistor (which would then be noiseless) with the level

$$\bar{e}^2 = 4 \cdot k \cdot T \cdot R \cdot B$$

The two methods are equivalent; however, the ''series voltage'' technique adds another node to the circuit. Thus, the ''parallel current'' technique is used.

For the entire circuit, each noise generator's contribution is calculated and propagated, by the appropriate transfer function, to the output of the circuit. At the output, all of these contributions are RMS-summed to obtain the ''total output noise.'' RMS stands for root mean square, that is, each contribution is squared, then the square root is taken of the average of all these amounts (this is the technique for adding variances). Also, since the transfer function from output to anywhere else in the circuit is known, PSpice will calculate the ''equivalent input noise.'' Detailed reports may, optionally, be generated showing the individual noise contributions.

## 8.2 THE .NOISE STATEMENT

The ''.NOISE'' statement directs PSpice to perform the noise calculations and specifies which nodes are the output, and where the input is. Moreover, the noise calculations are done over a range of frequencies and **the noise analysis is done in conjunction with a frequency response analysis** (.AC). This means that both a .NOISE and .AC statement are required to do the noise calculations. The .AC statement sets the frequencies at which the noise calculations are done. For the statement

.NOISE V(<*node*> [,<*node*>]) <*source name*> [*interval value*]

V(<*node*> [,<*node*>]) is the total output noise voltage: it may be a single node, in which case the noise voltage is referenced to ground, or a pair of nodes, in which case the noise voltage is taken to be across the two nodes. <*source name*> is the name of an independent source (V-device or I-device) to which the total output noise will be referred when calculating equivalent input noise. This source is not a noise generator; it is just a reference for describing the input (most likely it is the input for your frequency analysis).

If present, [*interval value*], causes the detailed printout of individual device noise contributions. As you may have guessed, this happens every *n*th frequency where *n* is the interval value. The individual contributions are referred to the output, so you may judge how important each is to the overall noise performance of the circuit, and are not the noise amounts for each contributor. The detailed printout, if specified, is generated regardless of any other output you might specify.

## 8.3 PRINT AND PLOT OUTPUT

Output from noise analysis may be generated by .PRINT or .PLOT statements, just as in AC analysis. In either case the output is organized by the frequency at which the calculations were made. The statement forms are

.PRINT NOISE <*output value*> . . .

and

.PLOT NOISE <*output value*> . . .

Each <*output value*> entry becomes a column in the table output by the .PRINT statement, or a curve in the plot output by the .PLOT statement. The output values you can print/plot are predefined as

| | |
|---|---|
| ONOISE | total noise at the designated output |
| INOISE | ONOISE referred to the input source |
| DB(ONOISE) | ONOISE in decibels (referred to 1-volt/hertz$^{1/2}$) |
| DB(INOISE) | INOISE in decibels (referred to 1-volt/hertz$^{1/2}$) |

## 8.4 GRAPHICS OUTPUT

Probe can also graph the values INOISE and ONOISE, and their values in decibels, which makes it easy to display results and make calculations. Let us take a look at a simple example of noise as demonstrated in a phonograph system to highlight the limits imposed by noise.

Most of the noise of an amplifying system is introduced in the first stage of the system, and, of course, once the noise is introduced nothing can be done to reduce its effects. In a phono-preamplifier, even if the amplification of the signal from a magnetic pickup were noise-free, there is still thermal noise generated by the magnetic cartridge, itself, and the resistive load for which it was designed. We may model the phono-cartridge and preamplifier as shown in Figure 8.1.

The amplifier is assumed to be noiseless. The section marked ''RIAA equaliza-

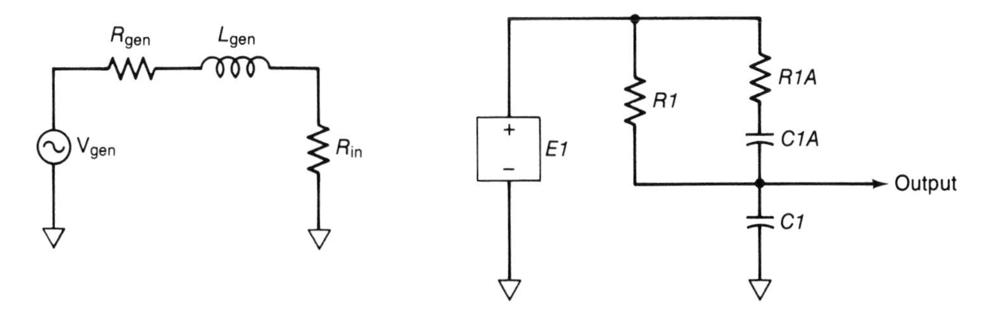


**Figure 8.1** Schematic of phono preamplifier circuit.

tion'' refers to a standard pre-emphasis that was involved in cutting the master record mold. This pre-emphasis was designed to limit the excursion of the groove at low frequencies and to improve signal strength at high frequencies. The preamplifier has to undo this pre-emphasis, with an opposite de-emphasis, and in doing so modifies the noise results from what you would have for an amplifier without de-emphasis. The circuit file for this follows:

```
Noise from magnetic phono-cartridge
Vgen 1 0 AC 1
Rgen 1 2 1350
Lgen 2 3 .5
Rin  3 0 47K
E1   4 0 3 0 10 ; first pole of RIAA curve @ 50Hz w/20db boost
R1   4 5 1
C1   5 0 3.528m
R1A  4 6 212.8m ; pole @ 2120Hz
C1A  6 5 352.8u ; zero @ 500Hz
.ac dec 100 20 20K
.noise v(5) Vgen 100
.probe
.end
```

You will notice that we have selected the output of the equalization as our noise output, and the signal generator of the phono-cartridge as our reference input. We have selected a frequency response analysis with a range of 20 hertz to 20,000 hertz to cover the entire audio range. This is the range over which the noise calculations will be made. After running PSpice, we may graph the noise results directly. But first, let's look at the transfer function of the equalizer, so you will understand some of the noise calculations we try later. The transfer function is displayed as shown in Figure 8.2.

With 1000 hertz as the unity gain reference frequency, you can see that the lower frequencies get quite a boost. Let us see what this does to the noise from our pre-amplifier.

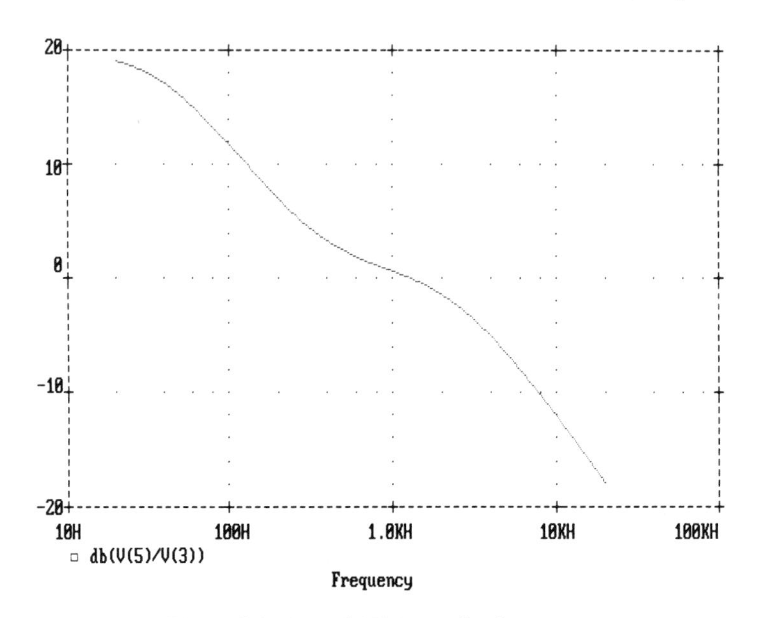


**Figure 8.2** Plot of RIAA equalization curve.

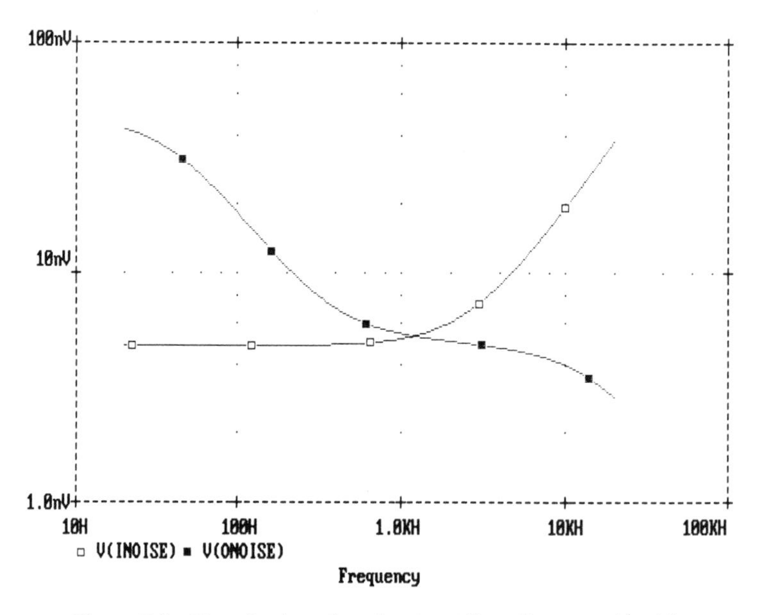


**Figure 8.3** Plot of noise referred to input (in volts per root-hertz).

The boost in the equalization circuitry amplifies the noise from the phono-cartridge and input load. Fortunately, the human ear is not as distracted by noise at lower frequencies as those in the mid-range (where the noise has already approached its lowest value).

## 8.5 CALCULATING TOTAL NOISE AND S/N

Using the phono-cartridge example from the previous section, we will use Probe to directly calculate noise totals. Total noise is the overall variance of the combined noise fluctuations at each frequency. This is the RMS calculation we discussed earlier. In Probe, so long as the frequency range for the calculation has been simulated, we can directly calculate total noise as shown.

The graph in Figure 8.4 is the running total of the noise contributions at each frequency, so that the right-most point on the graph is the total for the entire range of frequencies. The cursors are used, in Probe, to measure the total RMS noise across the band. The total noise is the Y value of the cursor. Although the greatest noise values occurred at low frequencies, we can see that the higher frequencies made up for this by having more bandwidth! Even so, nearly one-half of the total noise comes from the 20 hertz to 200 hertz band.

Now that we can calculate total noise, calculating signal to noise, or "S/N," is similar since

$$\text{S/N} = 20 \cdot log(\text{signal/total noise})$$

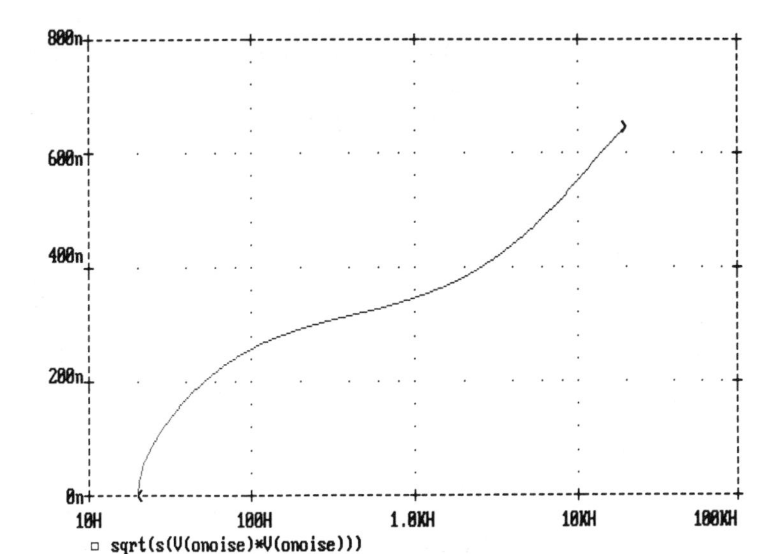


**Figure 8.4** Plot of total noise.

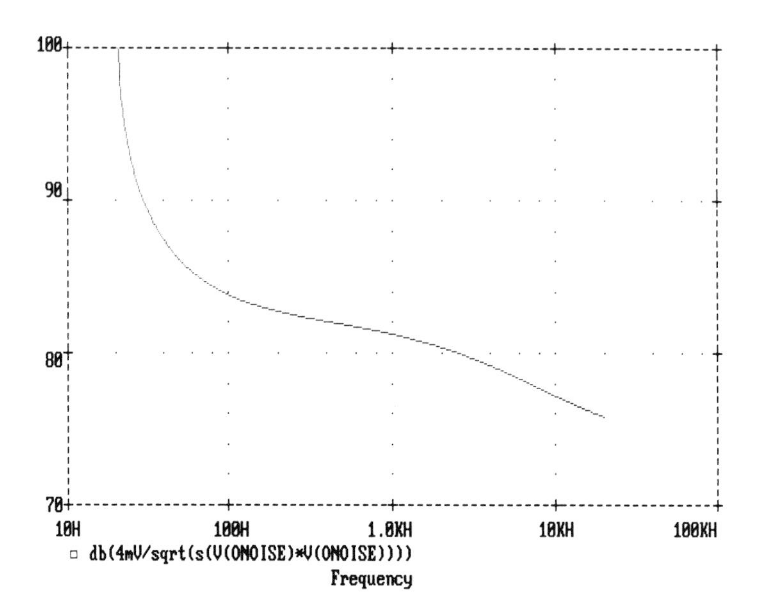


**Figure 8.5** Plot of signal-to-noise.

That is, value for S/N is the ratio of signal power to noise power expressed in decibels. Given that our magnetic phono-cartridge, in this example, generated an average signal (after de-emphasis) of 4mV, we may now display the graph for S/N.

Again, the graph in Figure 8.5 shows the running result for S/N, starting at 20 hertz. We can see that for the entire audio range, the limiting signal-to-noise ratio to expect from an ideal pre-amplifier is under 80 db.

## 8.6 INSERTING NOISE SOURCES

So far, we have considered only the noise generated by resistors. The active devices also have noise associated with their operation, which we will review in Chapter 12 on the semiconductor devices. But what if you want to insert a noise source directly? For example, you may want to enhance the modeling of an idealized circuit by providing for noise. This may be done by using the controlled sources to ''insert'' noise.

The noise analysis done by PSpice will calculate the value of the noise voltages, or currents, referenced to the input you have selected. This is a way of modeling noise for the entire circuit where the circuit is assumed to be noiseless, with all of the noise being generated by a noise source at the input to the circuit. Using this approach makes it easy to compare the merits of one circuit to another, when comparing noise specifications. You will see this in the data books, say, for operational amplifiers where the amplifier has a specified, nominal gain, as well as a specified input

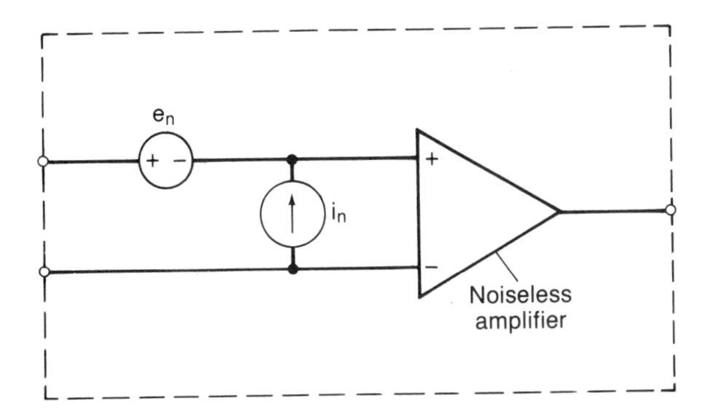


**Figure 8.6** Schematic of operational amplifier with noise sources.

noise voltage and current. Of course, the value for the noise sources vary with frequency, but for most work the constant values may be assumed. The diagram for an operational amplifier, with noise sources included, is shown in Figure 8.6.

The noise voltage is a voltage source in series with the input, since the operational amplifier is a voltage amplifier (that is, it has a high input impedance). The noise current is in parallel with the input and the effect of this noise source is determined by the resistance of the input circuitry (such as a transducer). In accordance with Ohm's law, the value of the noise current is converted to a noise voltage by the value of the resistance, or impedance. If the input to the amplifier has a low impedance, then the noise generated by the current noise source will add little to the overall input noise voltage. This would be the case if, say, the input to the amplifier were the output of another amplifier which has a fairly low output impedance. However, if the circuitry generating the input signal to the amplifier were of high impedance, such as a crystal microphone, then the input current noise will be much greater. The limiting value of the noise contributed by the noise current is determined by the input impedance of the amplifier.

Creating models for these noise sources is a simple matter of transferring the noise from a source, such as a resistor, to the spot in the circuit where you want it. From earlier in this chapter, we saw that the noise generated by a resistor, in PSpice, is

$$\bar{i}^2 = 4 \cdot k \cdot \mathrm{T} \cdot \mathrm{B/R} \qquad \text{for current noise}$$

$$\bar{e}^2 = 4 \cdot k \cdot \mathrm{T} \cdot \mathrm{R} \cdot \mathrm{B} \qquad \text{for voltage noise}$$

These noises may be measured by attaching the resistor to a zero-level source. Then the current noise would be current flowing through a zero-volt voltage source connected across the resistor (so long as the resistor and voltage source are isolated). Now that the noise current is flowing through a V-device, we may use one of the controlled sources, such as an F-device, to transfer or "insert" the noise into another part of the circuit.

As an example, let's generate a $1\mathrm{pA/Hz}^{1/2}$ as our "standard" and create a noise current of $5\mathrm{pA/Hz}^{1/2}$. At room temperature (300°K) the resistance required for the noise we want is

$$R = 4 \cdot k \cdot T/i^2 = 4 \cdot (1.67 \cdot 10^{-23})300/10^{-24} = 4 \cdot 1.67 \cdot 300 \cdot 10 = 20{,}040 \text{ ohms}$$

Then we connect this resistance across a zero-volt V-device, and transfer the noise current with an F-device that has a gain of 10, to arrive at a noise current level of $10\text{pA/Hz}^{1/2}$. The circuit file for this might look as shown:

```
Rnoise 1 0 20.04K
Vsense 1 0 DC 0
Fnoise 2 3 Vsense 10
```

Nodes 2 and 3 are the output of the new current noise. And, since it is a noise source, you don't need to worry about which way to connect it into the circuit; just connect it across the input to the operational amplifier.

For a noise voltage source, we could use an analogous technique of measuring the noise voltage across an isolated resistor, then transferring that noise with an E-device. However, we will be lazy by using the same circuit as before. This time we will generate a $1\text{nV/Hz}^{1/2}$ "standard" and insert a $3\text{nV/Hz}^{1/2}$ noise. First, calculate the resistance required for 1nA (yes, nano-amp, not nano-volt) of noise. Then we use an H-device, instead of an F-device, which will convert the noise current to a noise voltage. The circuit file for this might look as shown:

```
Rnoise 1 0 20.04m
Vsense 1 0 DC 0
Hnoise 2 3 Vsense 3
```

Nodes 2 and 3 are the output of the new voltage noise. Again, since this is a noise source, you won't need to worry about which way to connect it; just insert the H-device in series with the input to the operational amplifier.

These circuits provide the noise levels we need for modeling using "pure" noise (voltage and current) sources so as to not load the circuit at all. This follows the model for referring all noise to the input of the circuit. However, you will need to use a different resistor for each source you make, so that the generated noise is uncorrelated. You **should not** use just one resistor as the basis for all of your noise generators as they would be correlated (note that this is the way to create correlated noise, if that is what you want). It might be useful to create a set of noise generator subcircuits of standard levels used by your circuits.

**Exercise 8.6.1**

Develop a circuit that uses the measured noise across a resistor as the reference noise. Create both an independent noise voltage circuit and noise current circuit.

**Exercise 8.6.2**

Demonstrate the effects of correlated noise by using the same (generated) noise source twice in a circuit. Resimulate using two independent noise sources. Were the noise values what you expected?

# CHAPTER 9

# Transient Response

Transient, or time-domain, response is the most often used analysis for simulators like PSpice. This type of analysis attempts to simulate the operation of your circuit as time progresses and various inputs change level or as the circuit oscillates (because it is designed to oscillate) under the control of component values. Transient analysis is also the most trouble-prone analysis because of the compromises that need to be made to either (i) take small time-steps to ensure accuracy (but the simulations take a long time to complete), or (ii) take large time-steps with reduced accuracy (and possibly skip important features of the circuit response).

## 9.1 SIMULATING TIME

Without getting philosophical about "what time is," we observe that circuits behave predictably, and repeatably, with the progression of time. The changes in node voltages and branch currents are described by laws, and these descriptions are used by PSpice to simulate how a circuit will behave. We saw this with the DC sweep and AC response analyses, and it is not much different for transient analysis with the exception that to predict "forward" in time, the assumption is made that the currents, voltages, and element values will not change much from their present values. This is simplified, of course, but as an example you might consider that if, at the present time, an amount of current is flowing into a capacitor, then at the next moment very much the same amount of current will be flowing. If this is true, then the simulator may reliably predict the change in the voltage across the capacitor. If there is a big change in the two amounts of current then the simulator needs to take smaller time-steps. In the end, if the steps taken were small enough the calculations will approximate the circuit response for continuous time.

All of this is complicated by the use of active elements, which we have not covered so far (but you have some idea about their operation), that have regions of different operating characteristics. Diodes conduct readily in one direction of current, transistors cease amplifying with small changes in voltage, and so on. These gross changes in operation force the simulator to slow down and step carefully "around the curves" in the simulated response. Sometimes the numerics for doing this fail and produce chaotic results which halt the simulator (non-convergence at a time point). We will look at these problems.

## 9.2 SPECIFYING INPUT SOURCES

You recall that the independent voltage sources (V device) and current sources (I device) had the statement form

*<name> <node> <node> <value>*

where value was the DC or AC voltage or current level, depending on the device type. A fuller representation of the input source statement is

*<name> <node> <node> <DC value> <AC value> <transient value>*

where if you leave out the DC value, the DC value is set to zero, and likewise for the AC value. You would include values for all of the situations where you want to use an independent source that has a DC value, an AC value, and a transient value if you were doing all of these types of analyses. The DC value will be used for the operating point analysis and DC sweep. The AC value may combine with the DC value to set the operating point for the AC analysis. The transient value will override the other specifications **only** during the transient analysis. If the transient value is not specified, then the DC value will be used and the source is assumed to remain constant during the simulation.

The *<transient value>* portion of the statement has several forms, one for each type of waveform. If present, *<transient specification>* must be one of

| | |
|---|---|
| EXP *<parameters>* | for an exponential waveform |
| PULSE *<parameters>* | for a pulse waveform, which may repeat |
| PWL *<parameters>* | for a piecewise linear waveform |
| SFFM *<parameters>* | for a frequency-modulated waveform |
| SIN *<parameters>* | for a sine wave |

all of which are shown in Figures 9.1(a) through 9.1(e). Note: while all of the descriptions are for the independent voltage source, the same parameters are available for the independent current source (except that the units of volts are replaced with amps).

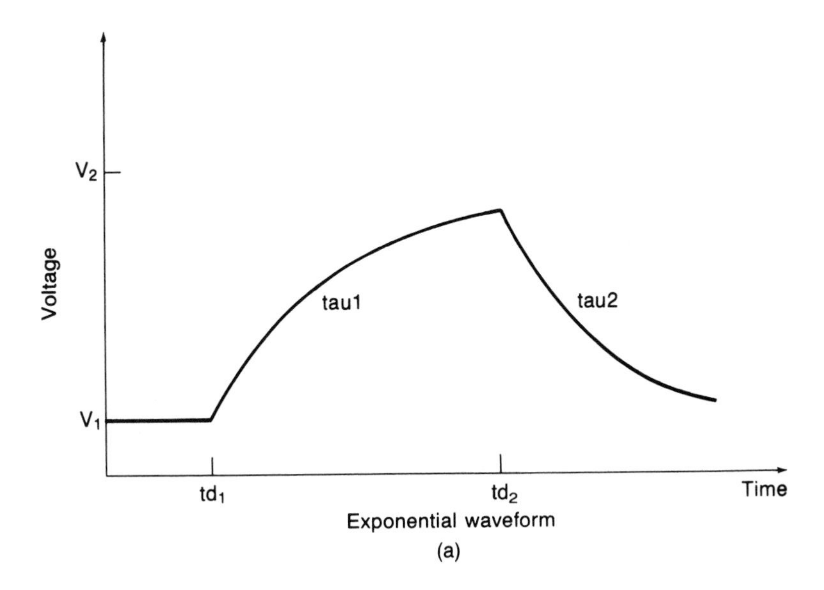


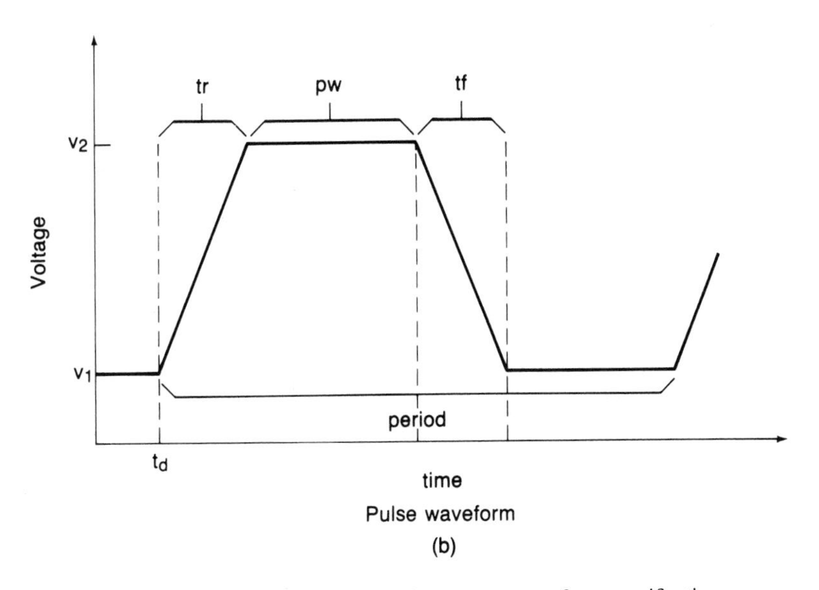


**Figure 9.1** Diagrams of independent source waveform specifications.

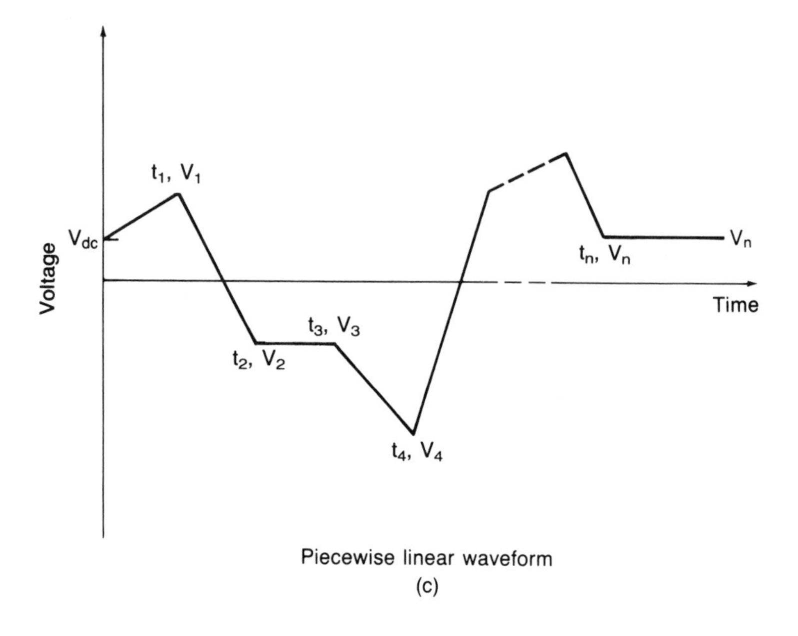


Piecewise linear waveform
(c)

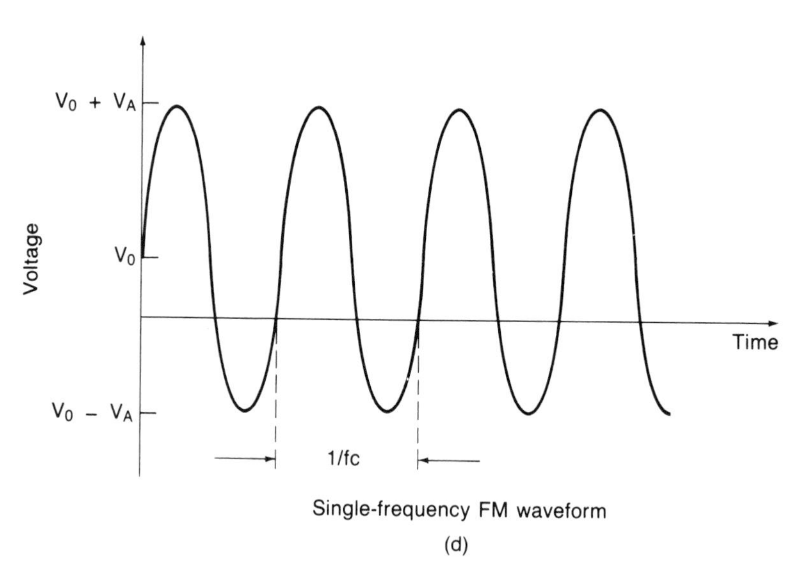


Single-frequency FM waveform
(d)

**Figure 9.1** *(Continued)*

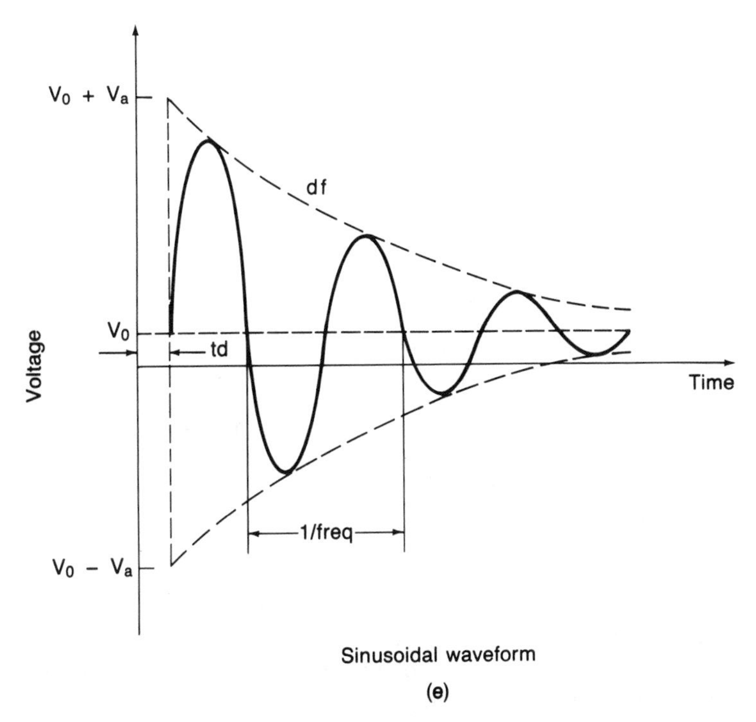


Sinusoidal waveform

(e)

**Figure 9.1** *(Continued)*

General Form

EXP (<*v1*> <*v2*> <*td1*> <*tau1*> <*td2*> <*tau2*>)

*Example*

```
VRAMP 10 5 EXP(0 .2 2uS 20uS 40uS 20uS)
```

| Parameters | | Default value | Units |
|---|---|---|---|
| <*v1*> | initial voltage | none | volt |
| <*v2*> | peak voltage | none | volt |
| <*td1*> | rise delay time | 0 | sec |
| <*tau1*> | rise time constant | TSTEP | sec |
| <*td2*> | fall delay time | <*td1*> + TSTEP | sec |
| <*tau2*> | fall time constant | TSTEP | sec |

The EXP form causes the voltage to be <*v1*> for the first <*td1*> seconds. Then, the voltage decays exponentially from <*v1*> to <*v2*> with a time constant of <*tau1*>. The decay lasts <*td2*> – <*td1*> seconds. Then, the voltage decays from <*v2*> back to <*v1*> with a time constant of <*tau2*>.

General Form

PULSE (*<v1> <v2> <td> <tr> <tf> <pw> <per>*)

*Example*

```
VSW 10 5 PULSE(-1V 1V 50uS .1uS .1uS 2uS 10uS)
```

| Parameters | | Default value | Units |
|---|---|---|---|
| *<v1>* | initial voltage | none | volt |
| *<v2>* | pulsed voltage | none | volt |
| *<td>* | delay time | 0 | sec |
| *<tr>* | rise time | TSTEP | sec |
| *<tf>* | fall time | TSTEP | sec |
| *<pw>* | pulse width | TSTOP | sec |
| *<per>* | period | TSTOP | sec |

The PULSE form causes the voltage to start at *<v1>* and stay there for *<td>* seconds. Then, the voltage goes linearly from *<v1>* to *<v2>* during the next *<tr>* seconds. Then, the voltage stays at *<v2>* for *<pw>* seconds. Then, it goes linearly from *<v2>* back to *<v1>* during the next *<tf>* seconds. It stays at *<v1>* for *<per>*-(*<tr>* + *<pw>* + *<tf>*) seconds, and then the cycle is repeated except for the initial delay of *<td>* seconds.

General Form

PWL (*<t1> <v1> <t2> <v2>* · · · *<tn> <vn>*)

*Example*

```
V3 10 5 PWL(0 - 1V 1uS 0V 10uS 0V 10.1uS 10V 20uS 10V 20.1uS 20V)
```

| Parameters | | Default value | Units |
|---|---|---|---|
| *<tn>* | time at corner | none | sec |
| *<vn>* | voltage at corner | none | volt |

The PWL form describes a piecewise linear waveform. Each pair of time-voltage values specifies a corner of the waveform. The voltage at times between corners is the linear interpolation of the voltages at the corners.

General Form

SFFM (<*vo*> <*va*> <*fc*> <*mdi*> <*fs*>)

*Example*

```
VFM 10 5 SFFM(0 2V 101MEGHz 5 4KHz)
```

| Parameters | | Default value | Units |
|---|---|---|---|
| <*voff*> | offset voltage | none | volt |
| <*vampl*> | peak amplitude of voltage | none | volt |
| <*fc*> | carrier frequency | 1/TSTOP | hertz |
| <*mdi*> | modulation index | 0 | |
| <*fs*> | signal frequency | 1/TSTOP | hertz |

The SFFM (Single-Frequency FM) form causes the voltage to follow this formula:

$$\text{vo} + \text{va} \cdot \sin(2\pi \cdot \text{fc} \cdot \text{TIME} + \text{mdi} \cdot \sin(2\pi \cdot \text{fs} \cdot \text{TIME}))$$

General Form

SIN (<*vo*> <*va*> <*freq*> <*td*> <*df*> <*phase*>)

*Example*

```
VSIG 10 5 SIN(0 .01 100KHz 1mS 1E4)
```

| Parameters | | Default value | Units |
|---|---|---|---|
| <*vo*> | offset voltage | none | volt |
| <*va*> | peak amplitude | none | volt |
| <*freq*> | frequency | 1/TSTOP | hertz |
| <*td*> | delay | 0 | sec |
| <*df*> | damping factor | 0 | $\text{sec}^{-1}$ |
| <*phase*> | phase | 0 | degree |

The SIN form causes the voltage to start at <*voff*> and stay there for <*td*> seconds. Then, the voltage becomes an exponentially damped sine wave described by this formula:

$$vo + va\cdot\sin(2\pi\cdot(freq\cdot(TIME - td) - phase/360))\cdot e^{-(TIME - td)\cdot df}$$

Note: the SIN waveform is **for transient analysis only.** It does not have any effect during AC analysis. To give a voltage a value during AC analysis use an AC specification; for example,

```
VAC 3 0 AC 1V
```

will have an amplitude of 1 volt during AC analysis and be zero during transient analysis, whereas

```
VTRAN 3 0 SIN(0 1V 1KHz)
```

will be the other around.

During the transient analysis, all of the independent sources that have a transient specification will be activated. The remaining independent sources will maintain the value of their DC specification, or zero if there is no DC specification.

## 9.3 THE .TRAN STATEMENT

The ".TRAN" statement specifies the time interval over which the transient analysis takes place. It also specifies some limits on the way PSpice does the analysis and when hard-copy output will be generated. The statement form is

TRAN[/OP] <*print interval*> <*final time*> [<*no-print interval*> [<*step ceiling*>]]

where the last two values are optional (that is, you may include either the third value, or both the third and fourth values). For the simulator, "time" always starts at zero and proceeds up to the value of <*final time*>. The "/OP" option, which stands for Operating Point, commands PSpice to print out the table of node voltages calculated from the bias-point calculation for the transient analysis. Normally these voltages would be the same as the bias-point calculation from the other analyses, such as frequency analysis, unless you specified some initial conditions which applied only to transient analysis. Then the bias-point is likely to be different, and "/OP" will save the node voltages in the output file.

The value for <*print interval*> specifies when hard-copy (PRINT and PLOT) output will be generated. The value for <*no-print interval*> will suspend hard-copy output until that amount of simulated time passes, so you will have only PRINT or PLOT output for the final stretch of the simulation.

The value for <*step ceiling*> specifies the maximum size of time step PSpice may take in working through the transient simulation. If the ceiling value is not specified, then PSpice uses 1/50th of the duration of the simulation, which is the maximum time-step size for any transient simulation.

## 9.4 PRINT AND PLOT OUTPUT

Output from transient analysis may be generated by .PRINT or .PLOT statements, just as in DC and AC analysis. In either case the output is organized by the time at which the calculations were made. The statement forms are

.PRINT TRAN <*output value*> . . .

and

.PLOT TRAN <*output value*> . . .

Each <*output value*> entry becomes a column in the table output by the .PRINT statement, or a curve in the plot output by the .PLOT statement. The output values you can print/plot are node voltages and device currents (which also means source currents, as a source is also a device).

| | |
|---|---|
| .PRINT TRAN V(7) | prints the voltage at node 7 |
| .PRINT TRAN I(R1) | prints the current through R1 |

You may print several values in one table, and mix voltages and currents; for example:

.PRINT TRAN V(3) I(R2)

Usually you will want to print the analysis time in the first column to simplify finding results in the table, so PSpice does this for you; the first column, which comes before the columns you specify, always contains the analysis time.

## 9.5 GRAPHICS OUTPUT AND CALCULATIONS

Using Probe with transient analysis is identical to what we have done before with DC and AC analysis; just include a .PROBE statement to the circuit file. Let's try simulating our LC-filter example with time-varying stimulus. Initially, we will use a step waveform to simulate overshoot and ringing. Recalling the LC-filter circuit from the frequency analysis section:

```
Four double-pole, low-pass, LC-filters
Vin 1 0 pwl(0,0 .1m,1 5m,1 5.1m,0)
*Q=.5
R1 1 2 200
L1 2 3 10mH
C1 3 0 1uF
*Q=1
R2 1 4 100
L2 4 5 10mH
c2 5 0 1uF
*Q=2
R3 1 6 50
L3 6 7 10mH
C3 7 0 1uF
*Q=4
R4 1 8 25
L4 8 9 10mH
C4 9 0 1uF
.TRAN 1m 10m
.PROBE
.END
```

You can see that a piecewise-linear source was specified for the step input. This step has a 0.1-millisecond transition and a nearly 5-millisecond duration. The transient analysis has been specified to last for 10 milliseconds.

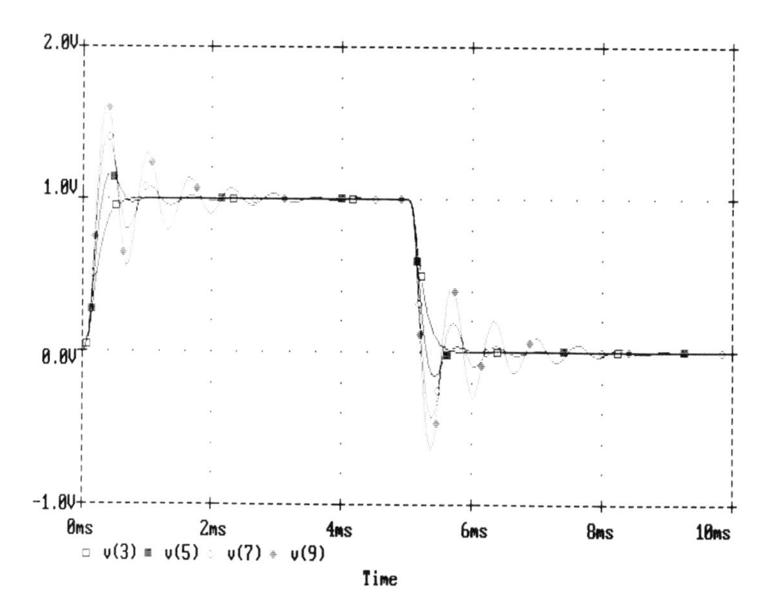


**Figure 9.2** Plot of LC-filter transient response.

Looking at the response of the filter sections, we see overshoot and ringing in the higher Q filters. Besides just looking at voltages and currents, with Probe, we can also make measurements using formulas.

In Figure 9.3 we see two derived measurements for (i) the volt-amp product, or instantaneous power dissipation, of resistor R4, and (ii) the running RMS average of this volt-amp product, or average power dissipation. If the step function were continued, as an oscillation, then the RMS average across one cycle would represent the long-term power dissipation of this resistor. In this case the dissipation would be slightly less than 0.3 milliwatts.

By using sinusoidal excitation, we can also see effects that were demonstrated using frequency analysis. To do this we modify the circuit file for the input source, as shown:

```
Vin 1 0 sin(0 1 2000)
```

which becomes a 2000 hertz sine wave with a 1-volt peak amplitude.

In Figure 9.4, we see different amplitudes of response to the same input. Since the input sine wave is slightly above the resonant frequency of the filter sections, the amplitude of response is set by the Q of the sections, with the higher Q sections having greater response. Also, the higher Q sections have greater phase lag. If you look closely, you will notice that the zero crossings of the higher Q sections come after those for the lower Q sections.

Finally, the step overshoot we saw previously is also apparent in the transient

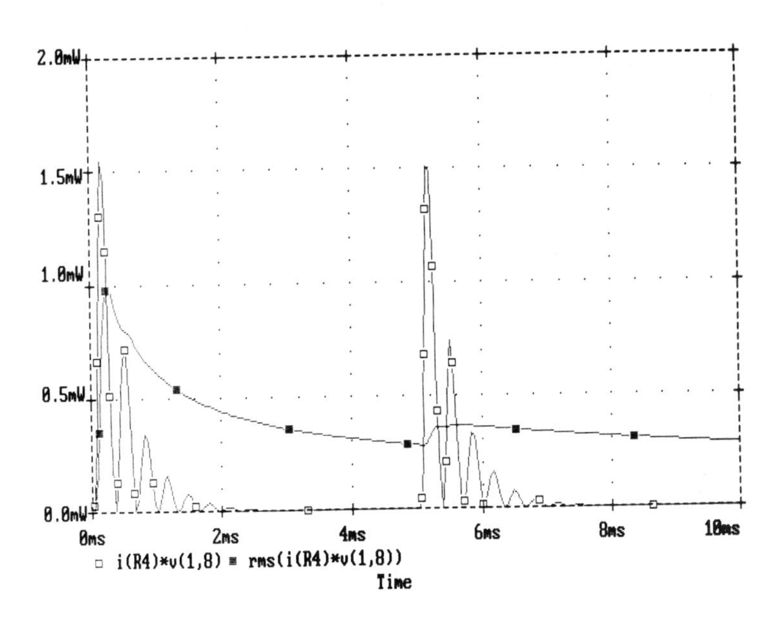


**Figure 9.3** Plot of instantaneous and average power.

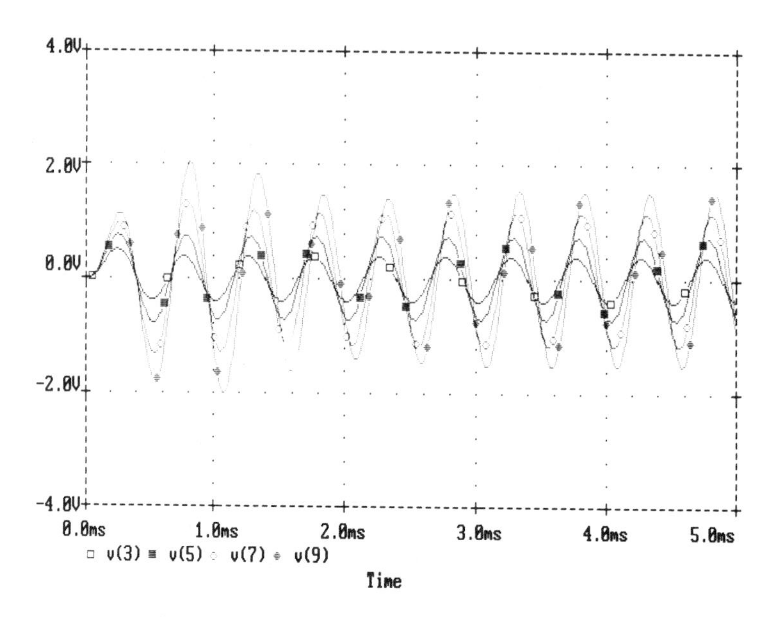


**Figure 9.4** Plot of sinusoidal excitation of LC-filter.

response to sinusoidal input. This initial transient is present because there is a change in the input, so that at TIME = 0 the input changes from nothing to a sine wave. This is a step change, and there is an overshoot and ringing period that dies out to become the steady-state response. This is one difference between transient and frequency response.

## 9.6 SETTING INITIAL CONDITIONS

For many of the simulations you run, you will want the circuit to begin with particular node voltages or device currents; for example, you may want to start with a particular inductor current. To do this you need to set the "initial conditions" for the simulation. There are two ways of setting initial conditions, and a related way of influencing the initial condition (for added confusion, which we will try to correct).

The first statement which directly sets the initial conditions is

.IC V(<*node*>) = <*voltage value*> . . .

which presents a list of nodes to be set to the indicated voltage. These statements are effective for **transient analysis only.** During the bias-point calculations, which is the setup for the transient analysis, PSpice connects voltage sources, each having a 2 milliohm source resistance, to the nodes specified by the .IC statements in your circuit file. This effectively fixes the nodes to those voltage levels. The 2 milliohm source resistance protects against voltage loops that are impossible to resolve. Then, just before the transient analysis begins, the voltage sources are removed and the circuit is initialized.

A related statement sets initial conditions for devices instead of nodes. Actually, it is not a statement but a part of the capacitor and inductor component statement. This augmentation of the statement was kept from you, until now, so you can see how it fits with the other methods of setting initial conditions. For the capacitor, you may set the initial voltage impressed on the "plates":

*capacitor statement* IC = <*initial value*>

For the inductor, you may set the initial current flowing through the windings:

*inductor statement* IC = <*initial value*>

These settings, like the .IC statement, are used for **transient analysis only.** Furthermore, the IC = option is used only when the .TRAN statement includes the "UIC" option, which stands for "Use IC =," as shown:

.TRAN [/OP] *list of time-values* [UIC]

This commands PSpice to skip the bias-point calculations and proceed to the transient analysis.

Setting the initial capacitor voltage this way lets you specify the relative voltages without having to specify the referenced-to-ground-voltages of the nodes of the capacitor. For the inductor, this is the only way to specify easily the current in the windings. Again, you do not need to specify the referenced-to-ground-voltages at the nodes of the inductor.

Of course, you can "over specify" the initial conditions by using both .IC and IC = to force, for example, conflicting voltage levels across a capacitor. Be careful.

The third, related way of setting up the simulation works during the bias-point calculation of all the analyses. This is the .NODESET statement, which is similar to the .IC statement, in form:

.NODESET V(<*node*>) = <*voltage value*> . . .

It **does not force the initial voltage at a node.** Rather, it provides PSpice with an initial guess at the outcome of the bias-point calculation and operates as follows:

first, voltages sources with source resistance of 2 milliohms (like the ones used for the .IC statement) are connected to the circuit where specified,

then the bias-point calculations are made and the circuit converges to a set of node voltages,

then the .NODESET voltage sources are removed, "releasing" the circuit,

then the bias-point calculations continue without the .NODESET voltage sources connected, and the final bias-point voltages are calculated.

As you can see, the .NODESET statement compels PSpice to use the specified voltages for only the first half of the bias-point calculation, then PSpice is free to recalculate the node voltages. This technique lets you give "hints" to PSpice about the bias-point voltages without fixing the nodes to a particular voltage. This is particularly useful for circuits that have more than one stable solution, especially balanced circuits (for example, flip-flops), which may have a metastable state that PSpice will be happy to use as the bias-point. The "hints" provided by the .NODESET statement may not even need to be that accurate, so long as they decide the issue of how you want the circuit to be biased.

To repeat, the .NODESET statement is used during the bias-point calculation for all of the analyses. Of course if you have both statements, .IC and .NODESET, specified for the same node the .IC statement will override during the bias-point calculation **for transient analysis only.**

## 9.7 HAZARDS: PROBLEMS OF TIME-STEPPED SOLUTIONS

The major problem of transient analysis is accuracy or (to be pessimistic) error. Since the circuit equations are solved numerically instead of analytically, approximations are used to extrapolate circuit operation at the next "instant" in time. Accuracy becomes a question of how good the approximations are and how far they may be extrapolated (see Appendix C for references explaining the operation of PSpice). Simulation accuracy is controlled by the parameter settings for RELTOL, VNTOL, ABSTOL, and CHGTOL, which are set by using the .OPTIONS statement (see Appendix A). The trade-off is computation time.

The foremost of the parameters is RELTOL, which sets the relative accuracy of the calculated voltages and currents. RELTOL is the numerical ratio of error allowed to the signal level; for example, a RELTOL value of 0.01 means that the voltages and currents are to be calculated to within 1 percent of their "real" values. But, how does the simulator determine what the "real" value is without calculating it? It doesn't . . . but the mathematical properties of the solution method let you estimate how close you are to the "real" value. So the "real" value, or answer, is within RELTOL of the answer calculated by PSpice. Just imagine a "band of uncertainty," with a width of

$$2 \cdot \text{RELTOL} \cdot \text{value}$$

centered on the waveform calculated by PSpice. Somewhere, in this band, lies the "real" waveform.

The other tolerance parameters, VNTOL, ABSTOL, and CHGTOL, set the best accuracy for the voltages, currents, and capacitor-charges/inductor-fluxes, respectively. That is, these parameters set the least error, or (optimistically) most accuracy, allowed in terms of an absolute amount. Now, why do we need this? Think about a waveform whose value changes sign. As it approaches zero, maintaining RELTOL accuracy will force the simulator to work harder because the tolerance, in absolute terms, is getting tighter. Ultimately, at zero, the uncertainty band containing the "real" waveform has zero width, and there is no assurance that the "real" value could ever be calculated. These tolerance parameters set the minimum error allowed, and the uncertainty band's width becomes, for a voltage, twice the maximum of RELTOL · value and VNTOL.

As the transient solution progresses, internal calculations are made for the next time point to evaluate the circuit. The size of each time-step is set as the minimum of several factors, but usually it is being set by a calculation of the errors involved in the integration techniques used for the energy storage elements (such as capacitors) in the circuit. If the node voltages are changing rapidly, then small time-steps are used to calculate accurately "around the curves." If the circuit becomes less active, then the node voltages are more stable and larger time-steps may be used.

Using ever larger time-steps becomes a problem when the step size exceeds the Nyquist rate for the "real" signals in your circuit. After all, the simulator is like a sampled-data system with samples being made at every time-step. If the samples become too widely spaced, then high-frequency operation will be "aliased" to a lower frequency. For example, an astable multivibrator circuit may be incorrectly simulated by allowing the simulator to have a step-size which is larger than one-half the oscillation period. In this case, the standard capacitor charge/discharge circuitry is never activated because the simulator has found a "stable" solution (a lie) where the discharge threshold is never reached.

To prevent the time-step from becoming too large and missing changes in the circuit, PSpice forces a time-step at each corner in the driving waveforms; for example the "pulse" and "piecewise linear" specifications. PSpice also limits the time-step to one-eighth of the cycle time of the sinusoidal source with the highest frequency. Another limit is twice the previous time-step size, which has proven to be a good, conservative measure for not allowing the simulator to get too far ahead and into trouble. In addition, you may also set the maximum step-size by using the fourth parameter of the .TRAN statement. PSpice uses the minimum of all the limits described to limit the size of the time-step.

## 9.8 BENEFITS: TRANSIENT SOLUTIONS FOR STATIC PROBLEMS

Transient analysis can provide some solutions to problems that you would ordinarily consider as static problems. The classic example is analyzing the transfer characteristics of a circuit with regenerative feedback, such as a Schmitt trigger. These types of circuits cannot be analyzed easily using a DC sweep. The circuit has a region with two stable operating points, and the simulator will need to jump discontinuously from one stable solution to the other at each end of this region.

By using transient analysis and the piecewise linear source, you can apply a slowly varying ramp to the same circuit, just as you might with the DC sweep. The difference here is that the circuit is discontinuous only in the sense of its DC operating point; in transient analysis the circuit is continuous, although at some threshold it does transfer rapidly from one operating point to the other. When the circuit switches like this, PSpice reduces its time-step size to analyze carefully the transition. During the remainder of the time the time-steps become quite large because the node voltages do not change so much.

This kind of treatment duplicates exactly what you would do on the workbench; you would slowly change the input and measure the output. Also, you would probably sweep the input in one direction, until the circuit switches, and then sweep in the other direction, until the circuit switches back. Doing this measures the hysteresis of the circuit. This is easy to do with the piecewise linear source in one transient run, but with DC analysis you would need to do a run for each direction of the sweep.

We can also use a rapidly varying ramp to measure the size of energy storage elements in a circuit, or device. As we reviewed earlier, a linear capacitor stores charge according to the equation

$$Q = C \cdot V$$

Differentiating this equation with respect to time, we get

$$dQ/dt = I = C \cdot dV/dt + V \cdot dC/dt$$

which simplifies to

$$I = C \cdot dV/dt$$

since: (i) if the capacitor is linear, then $dC$ is zero, and (ii) if the capacitor is nonlinear, then the original equation is wrong. In the latter case, charge is the integral of the capacitance, with respect to voltage; using integration by parts, with both voltage and capacitance as functions of time, before differentiating (the equation with respect to time) leads to the same result. Minor rearranging yields

$$C = I/(dV/dt)$$

And, of course, similar analysis for inductors will yield the equation

$$L = V/(dI/dt)$$

Let's get back to transient analysis techniques. We now have a way to measure the capacitance of a network versus voltage. By applying a voltage ramp that produces currents due to capacitance which are much larger than the DC currents (for the same voltage levels), the current represents the value of the capacitance along the ramp. Or, we could measure inductance versus current by using a current ramp. Just be sure to isolate and measure one energy storage element at a time. (Later in the book we will see this technique used to plot the capacitance versus voltage characteristics of the semiconductor devices.)

## 9.9 UNUSUAL WAVEFORM SOURCES

Sometimes your simulations will need a stimulus which is more exotic than a simple sine wave or piecewise linear function. With some ingenuity, you will find that you can create many unusual waveforms by combining the ones available through the use of controlled sources. For example, to have a 10Khz signal riding on a 60Hz power line you might use the voltage-controlled voltage source:

```
vsig 1 0 sin(0 1 10k)
rsig 1 0 1
vpwr 2 0 sin(0 120 60)
rpwr 2 0 1
eboth 3 0 poly(2) 1,0 2,0 0 1 1
```

although you could have added these signals by putting the sources in series.

Multiplying, or modulating, signals is another way to generate desired inputs. For example, to create a ten-cycle burst of sine waves you might multiply a sine wave source with a piecewise "switching" function:

```
vsig 1 0 sin(0 1 1K)
rsig 1 0 1
vsw  2 0 pwl(0,0 .001m,1 9.999m,1 10m,0)
rsw  2 0 1
eboth 3 0 poly(2) 1,0 2,0 0 0 0 0 1
```

which will produce ten cycles of a 1KHz sine wave at the start of the simulation, and then be zero for the remainder of the run. This might be used to simulate the reaction of a filter circuit. You may notice that, in this example, a .001-millisecond transition time was used for the switching function. This is much shorter than the 1-millisecond cycle time of the sine wave, so it should not deviate much from a true sinusoidal shape. However, the transition should not be too short as PSpice will cut back on its time-step to process the changes in the switching function. Then it will take some time to get moving again with the simulation.

By extension, you could build several of these generators, with interleaved "on" times, to step through a series of frequencies. You have to be careful that the zero crossings of the sine waves occur when the switching functions transfer from one source to another, to get a clean transition. You may need to adjust the initial phase of a source to make this happen. Of course, it is a good idea to do a run with just the sources to check their operation.

Amplitude modulation may be done in a similar way. For example, modulating a 500KHz signal, at an 80 percent modulation index, with a 1KHz sine wave:

```
vsig 1 0 sin(0 1 500K)
rsig 1 0 1
vmod 2 0 sin(1 .8 1K)
rmod 2 0 1
eboth 3 0 poly(2) 1,0 2,0 0 0 0 0 1
```

Since we are doing a simple multiplication of the two signals, the modulation index is the ratio of the peak amplitude of the modulating signal to unity. For the same reason the modulating signal has an offset of 1 volt, or unity, so that if the carrier signal were unmodulated its peak amplitude will be 1 volt.

# CHAPTER 10

# Distortion and Spectral Analysis

Signal distortions come in many forms, most of them undesirable, and are usually the product of nonlinearity in the gain, or nonuniformity in the phase, of a circuit. The most common types of distortion have been categorized and named:

*Harmonic distortion* comes from nonlinear gain. The output of the circuit contains integral multiples, or harmonics, of the fundamental input frequencies.

*Phase distortion* comes from nonlinear phase versus frequency response. This gives rise to ''echoes'' in the output that precede and follow the main response, resulting in a distortion of the output signal when the input signal has many frequency components.

*Intermodulation distortion* comes from mixing signals at different frequencies. The output of the circuit contains signals at integral multiples of the sum or difference of the original frequencies.

*Cross-modulation distortion* occurs when the modulation of one signal is unintentionally transferred to another signal in the circuit.

*Crossover distortion* comes from nonlinearities in amplification as the signal crosses over between regions of amplifier operation (such as a ''push-pull'' amplifier).

## 10.1 THE .DISTO ANALYSIS

The SPICE2 simulator, from U.C. Berkeley, has a type of analysis called .DISTO for evaluating some distortion measures. This analysis is performed in conjunction with a frequency analysis, the way noise analysis is done, that is calculations are performed at the frequencies specified by the .AC statement at the time the frequency analysis is done. The .DISTO analysis calculates the magnitude of power into a load resistor for the following small-signal harmonic products:

the power of $2 \cdot f$
the power at $3 \cdot f$

where $f$ is each frequency specified as part of the .AC statement, and for the following small-signal intermodulation products:

the power at $f + \mathrm{k} \cdot f$
the power at $f - \mathrm{k} \cdot f$
the power at $2 \cdot f - \mathrm{k} \cdot f$

where the factor k is specified by the .DISTO statement. The relative magnitude of the second signal, represented by $\mathrm{k} \cdot f$, may also be set by the .DISTO statement. The results of the .DISTO analysis are available to the .PRINT and .PLOT output statements.

PSpice does not include the .DISTO statement because:

The calculations are for small-signal distortion (only) whereas many of the interesting distortion analyses are for large signals.

The calculations are for power at only a few, selected frequencies, whereas most circuits exhibit distortion components at a large number of frequencies.

Moreover, the results were incorrect, particularly for MOS circuits, where the small-signal model equations had not kept up with advances in the nonlinear (large-signal) model.

Instead, PSpice makes use of spectral analysis techniques to calculate distortion.

## 10.2 HARMONIC (FOURIER) DECOMPOSITION

One type of spectral analysis is part of the PSpice simulator. It is called *Fourier analysis*, and is done in conjunction with the transient analysis. Jean Fourier (1768–1830) developed mathematics demonstrating that any periodic function could be expressed as the sum, or series, of sinusoidal functions. Moreover, if a periodic function is expressed this way, then each sinusoid, or "component," in the series must be periodic over the same interval as the original function. This happens only with sinusoids having frequencies that are an integer multiple of the frequency of the original function; that is, the sinusoids are "harmonics" of original function's "fundamental" frequency.

The .FOUR statement directs the simulator to perform a "harmonic decomposition," calculating the "Fourier coefficients" for the sinusoidal components of any voltage or current that you could PRINT or PLOT. Also, you select the fundamental frequency on which to base the decomposition. These calculations create tabulated results which include the DC component, the fundamental component, and the components of the second through ninth harmonic of the fundamental. The magnitude and phase values, both absolute and relative to the fundamental, are compiled in the table. Finally, harmonic distortion is calculated.

It is important to remember that when a harmonic decomposition is taken of a transient waveform **only part of the waveform is used for the decomposition.** The period of time used is inverse of the frequency, or one cycle's time, of the fundamental frequency you specify for the decomposition. The segment of the waveform that is used is the last period (= 1/frequency) of the transient simulation. You will want to set up your simulations so that the segment of waveform decomposed is at the end of the simulation (usually you will just set the time limit for the entire transient run to be one cycle's time of the fundamental).

Finally, a note about harmonic distortion, which is defined to be the ratio of the root-mean-square (rms) sum of the magnitude of the harmonics to the magnitude of the fundamental, or as a percentage

$$\%\ \text{harmonic distortion} = 100 \cdot (V_2^2 + V_3^2 + V_4^2 + \cdots)^{1/2}/V_1$$

where $V_n^2$ is the magnitude of the *n*th harmonic, squared. Calculating the popular distortion measure called "percent total harmonic distortion" or "%THD" is straightforward, and is done for you by PSpice. Also, while setting up the transient analysis you can select the signal level, for example, to include the effects of crossover and clipping.

## 10.3 THE .FOUR STATEMENT

Remember, you have to be doing a transient simulation to get a harmonic decomposition using the .FOUR statement. Having covered that, we now look at the .FOUR statement and find that it is similar to a .PRINT statement, with the form

.FOUR *<fundamental frequency value>* *<output value>* . . .

One, or more, node voltages and/or device currents may be selected for harmonic decomposition. You have to choose the fundamental frequency for this analysis. Let's take a look at a small circuit which demonstrates Fourier analysis.

```
*Fourier decomposition
vin 1 0 sin(0 .57 1000)
rin 1 0 1G
```

```
e3  2 0 poly(1) (1,0) 0 1 0 -1
r3  2 0 1G
.tran 1u 1m
.four 1000 v(1) v(2)
.probe
.end
```

This circuit does a transient analysis of a 1000 hertz sine wave exciting a VCVS with the transfer function $x - x^3$. This is a cubic polynomial that was selected to demonstrate the distortion effects of nonlinear gain and "soft" clipping, since the input waveform has a peak value of nearly $1/3^{1/2}$, which is the point at which the cubic polynomial reverses direction. The transient simulation is run for one cycle of the input sine wave, 1 millisecond, since this is the fundamental waveform of the simulation. Note that both the input and output voltages were selected for decomposition. Seeing the decomposition of a known waveform will provide some guidance in interpreting the results.

The .PROBE statement was included so we could look at the waveforms; it is not required, and neither is any other output statement, like .PRINT or .PLOT, for the Fourier decomposition.

## 10.4 LARGE-SIGNAL DISTORTION

Running the circuit described previously will yield results not obtainable using the .DISTO analysis; since the DC bias-point has the input voltage at zero, the linear term of the transfer polynomial will dominate and the output will show very little harmonic distortion. Looking at the output file, we find two tables. The first is for the input voltage, V(1):

```
FOURIER COMPONENTS OF TRANSIENT RESPONSE V(1)
DC COMPONENT = 2.017000E - 10
```

| HARMONIC NO | FREQUENCY (HZ) | FOURIER COMPONENT | NORMALIZED COMPONENT | PHASE (DEG) | NORMALIZED PHASE (DEG) |
|---|---|---|---|---|---|
| 1 | 1.000E+03 | 5.700E-01 | 1.000E+00 | 3.062E-07 | 0.000E+00 |
| 2 | 2.000E+03 | 3.719E-10 | 6.525E-10 | 1.297E+02 | 1.297E+02 |
| 3 | 3.000E+03 | 1.568E-09 | 2.750E-09 | 8.906E+01 | 8.906E+01 |
| 4 | 4.000E+03 | 5.704E-10 | 1.001E-09 | 1.554E+02 | 1.554E+02 |
| 5 | 5.000E+03 | 9.842E-10 | 1.727E-09 | -2.179E+01 | -2.179E+01 |
| 6 | 6.000E+03 | 5.810E-10 | 1.019E-09 | 1.216E+02 | 1.216E+02 |
| 7 | 7.000E+03 | 2.297E-09 | 4.030E-09 | 1.311E+02 | 1.311E+02 |
| 8 | 8.000E+03 | 7.364E-10 | 1.292E-09 | 3.312E+01 | 3.312E+01 |
| 9 | 9.000E+03 | 9.024E-10 | 1.583E-09 | -1.565E+02 | -1.565E+02 |

```
TOTAL HARMONIC DISTORTION = 5.781659E-07 PERCENT
```

and a similar table for the output voltage, V(2):

```
FOURIER COMPONENTS OF TRANSIENT RESPONSE V(2)
DC COMPONENT=1.790805E-10
```

| HARMONIC NO | FREQUENCY (HZ) | FOURIER COMPONENT | NORMALIZED COMPONENT | PHASE (DEG) | NORMALIZED PHASE (DEG) |
|---|---|---|---|---|---|
| 1 | 1.000E+03 | 4.311E-01 | 1.000E+00 | 3.707E-07 | 0.000E+00 |
| 2 | 2.000E+03 | 4.006E-10 | 9.292E-10 | 1.282E+02 | 1.282E+02 |
| 3 | 3.000E+03 | 4.630E-02 | 1.074E-01 | 7.127E-06 | 6.756E-06 |
| 4 | 4.000E+03 | 5.130E-10 | 1.190E-09 | 1.408E+02 | 1.408E+02 |
| 5 | 5.000E+03 | 1.374E-09 | 3.187E-09 | 1.273E+02 | 1.273E+02 |
| 6 | 6.000E+03 | 4.424E-10 | 1.026E-09 | 1.073E+02 | 1.073E+02 |
| 7 | 7.000E+03 | 1.007E-10 | 2.336E-10 | 1.263E+02 | 1.263E+02 |
| 8 | 8.000E+03 | 5.842E-10 | 1.355E-09 | 4.881E+01 | 4.881E+01 |
| 9 | 9.000E+03 | 1.106E-09 | 2.566E-09 | 1.530E+02 | 1.530E+02 |

```
TOTAL HARMONIC DISTORTION = 1.073909E+01 PERCENT
```

For each of the selected voltages, the tables show the magnitude and phase of the components at 2x, 3x, . . . 9x the fundamental frequency. These are the harmonics of the fundamental. The values are also shown in a normalized format so that the magnitudes are normalized with the fundamental at unity, and the phases are normalized with the fundamental at zero.

Above the table, the DC component of the waveform is shown. Below the table, the calculation for ''total harmonic distortion'' is shown (of course, the calculation uses only the second through ninth harmonics).

Looking at the decomposition table for the input sine wave, we can see the limits of the calculations—this is a sort of ''noise floor'' due to the numerics. The input ideally has a zero DC level, and zero magnitude for all harmonics resulting in zero harmonic distortion. Instead, PSpice arrived at values of around one part in a billion (relative to the fundamental) . . . numerical noise. From this we can see that only those values which are larger than this ''noise floor'' are significant (and useful) values.

The same analysis applies to the phase values, so that as harmonic magnitudes become small it becomes difficult to determine phase. Again, this is a numerical ''noise'' problem. For simple systems, where the output is a periodic waveform, you know from Fourier analysis that the harmonics are either in phase (0 degrees) or out of phase (180 degrees). So, sometimes you have to apply some judgment in reading the tables.

Looking at the decomposition table for the output waveform, we can see that there is one major harmonic at three times the frequency of the fundamental. It is in phase and has a level about 11 percent of the fundamental. Accordingly, the total harmonic distortion is also calculated at about 11 percent.

**Exericse 10.4.1**

Show that the .FOUR results will yield smaller distortion values by reducing the size of the input sine wave. Try .1 volts peak and .01 volts peak.

## 10.5 HARMONIC RECOMPOSITION

One of the uses of the decomposition table printed by the Fourier analysis is providing the harmonic information for regenerating the signal that was decomposed. This may provide you with a compact and fast technique for packaging a complex waveform. Of course, it needs to be a periodic waveform that can be represented adequately with the first nine harmonics.

Looking at the previous table for the cubic transfer function, we find that the output waveform can be represented by summing two sine waves: one is the fundamental, with a magnitude of 0.4311, the other is at 3x the fundamental frequency, with a magnitude of 0.0463. We can combine these signals in the original circuit and compare our "recomposed" signal to the original. This is done by adding two current sources, which sum to develop a voltage across a 1-ohm resistor, to the circuit:

```
*Fourier decomposition
vin 1 0 sin(0 .57 1000)
rin 1 0 1G
e3  2 0 poly(1) (1,0) 0 1 0 -1
r3  2 0 1G
ix1 0 3 sin(0 .4311 1000)
ix3 0 3 sin(0 .0463 3000)
rx  3 0 1
.tran 1u 1m
.four 1000 (v(1) v(2) v(3)
.probe
.end
```

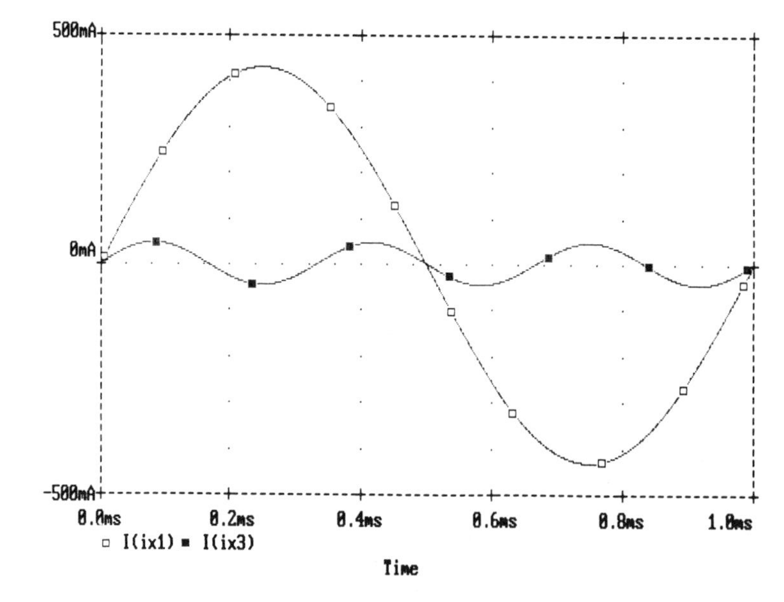


**Figure 10.1** Plot of fundamental and harmonic.

Looking at the output with Probe, see Fig. 10.1, we can see the components of our "recomposed" waveform. We can also check the accuracy of our work by checking the difference between the original waveform and the "recomposed" signal.

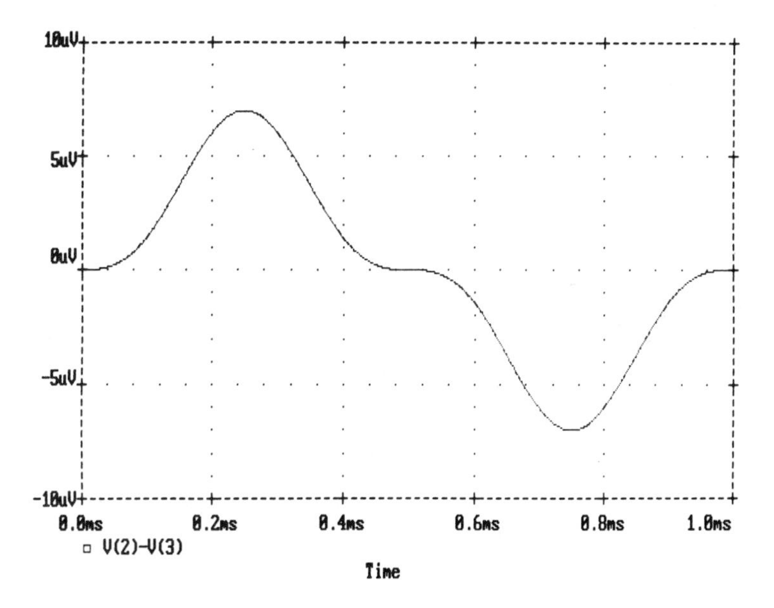


**Figure 10.2** Plot of signal error.

Notice the scale for the Y-axis, showing that the difference is quite small relative to the magnitude of the waveforms being compared.

**Exercise 10.5.1**

Do the same analysis for a circuit with the polynomial function $x - x^5$. What maximum input level will you use to show "soft clipping"? Look at the harmonics and compare their magnitudes with the ones in the example.

## 10.6 FOURIER TRANSFORM

Another type of Fourier analysis is part of Probe, which has the capability to calculate "Fourier transforms" of the data sequences in the Probe data file. While the Fourier transform is a special case of the Laplace transform, the Fourier integral (the function performing the transform) can also be thought of as an extension of the Fourier series we saw in the previous section; by extending the fundamental period to infinity each harmonic component becomes infinitesimally close in "frequency," so that in the limit the Fourier series becomes the Fourier integral. This means the Fourier transform converts a function of time to a function of frequency, and vice versa. The physical interpretation of Fourier transform is the conversion of a time-domain signal to the AC steady-state frequency content, or spectrum, which makes up the signal. However, the inverse Fourier transform converts the AC steady-state response, or gain, of a system into the time-domain response of that system to a flat input spectrum (which is an impulse).

The Fourier transform in Probe is a "discrete Fourier transform" (DFT), where the Fourier integral has been replaced by a nearly equivalent summation formula applied to evenly spaced samples of the signal. Furthermore, the transform is accomplished by a special technique credited to Cooley and Tukey, which is commonly called a "fast Fourier transform" (FFT). This is a numerical trick whereby if the size of the data sequence is a power of two (such as 256, or 4096) a much shorter sequence of calculations can be used to get the same results as the discrete Fourier transform. Even for modest data sets the DFT is so time consuming that most computer applications use the FFT (for example, for 1024 points only about 1 percent of the compute time is required using the FFT versus the "brute force" DFT). Probe uses the FFT, and gets its data by interpolating the data file values to get a set of values with the appropiate number of data points. Then the FFT is performed.

For Probe, the Fourier transform is a mode; all displayed values or formulas are transformed before being displayed (we will use the word "signal" to also mean a formula of signals). As you might expect, there are some items to be cautious about when using Fourier transforms:

> If the signal is nonzero for a finite interval, then you should transform that entire interval. Alternatively, if the signal goes on forever, then transform an interval that is "typical" of what the signal looks like at the other times.
>
> The transform should be made on a *band-limited* signal. The usual warnings about data sampling at greater than the Nyquist critical rate are taken care of by PSpice and Probe, since the simulation itself is subject to the Nyquist sampling theorem and the time-steps in the transient simulation are usually much smaller than the highest frequency signal. Probe uses the total number of time-steps to guide the FFT. However, if there is a difference in the value between the beginning and end of the waveform, this is a discontinuity which implies a DC level and/or a high frequency content which will be aliased (frequency shifted) into the result from the FFT. Fortunately, this will be evident when you look at the transformed signal, so you may need to retry the simulation for a different interval.

Let's use a variation on the circuit for the Fourier decomposition example to demonstrate Fourier transforms in Probe.

```
*Fourier transform example
i1 0 1 sin(0 .57 1000)
rin 1 0 1
e3 2 0 poly(1) (1,0) 0 1 0 -1
```

```
r3 2 0 1G
.tran 50u 10m 0 50u
.probe
.end
```

This circuit is the same as before, except that a current source is used to create the input voltage. This will make it easy to add in other signals, say, to show intermodulation effects. The transient simulation is run for 10 milliseconds, or 10 cycles, so that the frequency resolution will be 1/10 of the fundamental frequency. The time step is limited to 50 microseconds, or 1/20 of the cycle time, so that the frequency band, from zero up to the Nyquist critical frequency, will extend to about 10 times the fundamental frequency (actually set by the number of data points selected by Probe, but this will always be at least as large as the number of points in the waveform data). By running the simulation and then looking at the results with Probe, we see the input voltage in the time domain (see Fig. 10.3).

By selecting the X-axis menu and Fourier menu item, we can look at (after some delay for calculation) the signal's spectrum in volts per root hertz, since the signal is in volts (see Fig. 10.4).

Some comments are required here to help you interpret the results of the transform:

The total power (or mean-square amplitude) of the signal is the same whether it is represented as a time-domain function or transformed to be a spectral-domain function (this is a result of Parseval's theorem). For example, if you want to know how much power there is between two frequencies then you

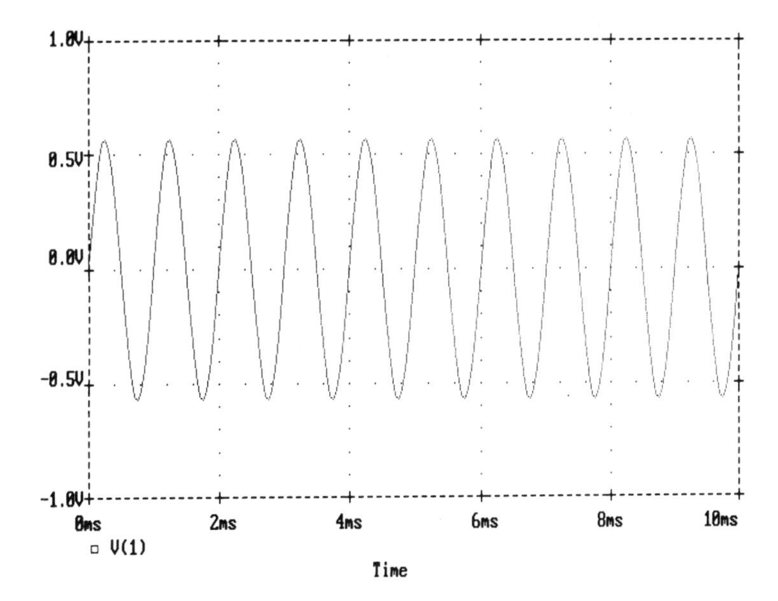


**Figure 10.3** Plot of input signal (time domain).

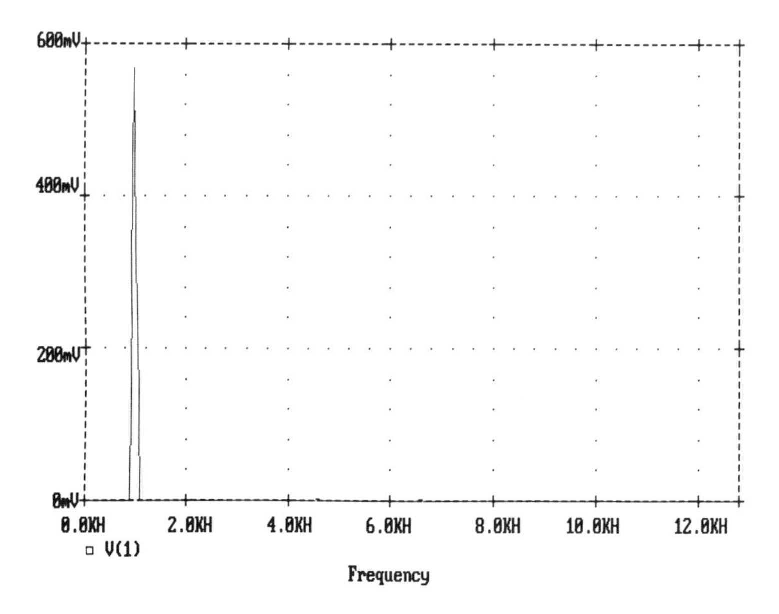


**Figure 10.4** Plot of input signal (spectral domain).

integrate the square of the spectral amplitude between those two frequencies.

The result from a discrete Fourier transform, such as the FFT, is **not the same** as the continuous Fourier transform values sampled at the same frequencies. Due to the lack of resolution of the DFT, each data point of the transform is actually the level of a "bin" (in frequency, for the transform of a time value) which extends halfway to the adjacent data points, with an area approximately the value of the area under the Fourier transform for the same range of frequencies.

Accordingly, the Fourier transform of a sine wave results in a delta function, that is, an infinitely high "spike" of zero width, at the frequency of the sine wave, whose area represents the amplitude of the sine wave. For the DFT, or FFT, of the same sine wave, the "spike" has finite height and non-zero width, but the area is approximately the same. If more samples are used for the DFT then the transform has more resolution (it becomes more like the continuous Fourier transform) and the DFT "spike" gets taller and narrower, its area converging to the value of the delta function.

For other sections of the spectrum (those not containing single frequency signals) you may as well consider the DFT, or FFT, result to be the sampled value of the Fourier transform at that frequency.

These caveats are not unique to discrete Fourier transforms; similar precautions apply to the use of some types of spectrum analyzer equipment.

However, the waveforms that result from computer simulations generally do not have continuous spectra. This is because the stimulating waveforms are generally pure, such as a sinusoidal waveform, or have a definite set of harmonics, such as a pulse. There is no broadband source, for transient response, available in the simulator. And since nonlinear circuits can convert only single frequency combinations into other single frequency combinations, the output waveforms will always contain only individual and distinct frequencies (like spectral lines, to use the analogy to light). Any response between these frequencies, as calculated by the FFT, is **due to numerical noise** in the interpolation and transform operations. So, we ignore the results of the FFT in these areas.

Now, since we are going to be looking at only the ''spikes'' in the transform results, there is no sense in manually calculating the area of these to find the signal power of each. The height of the ''spike'' is proportional to the number of samples going into the transform, so Probe scales the result for you. This means the height of the ''spike'' is really the calculated spectral amplitude, and this **will not vary** with the number of samples put into the FFT. For our example, the height of the ''spike'' is 0.5676, which is within 0.5 percent of the amplitude we set for the sine wave in this example.

Looking at the transform of v(2) we see the third harmonic ''spectral line'' (see Fig. 10.5). Measuring the ''spike'' at 1000 hertz we get 0.4293 which is, again, within 0.5 percent of the value we obtained using the harmonic decomposition (.FOUR statement). For the ''spike'' at 3000 hertz we get 43.16E-3, which is low by 7 percent from the value obtained using the harmonic decomposition. This error may be improved by using more samples.

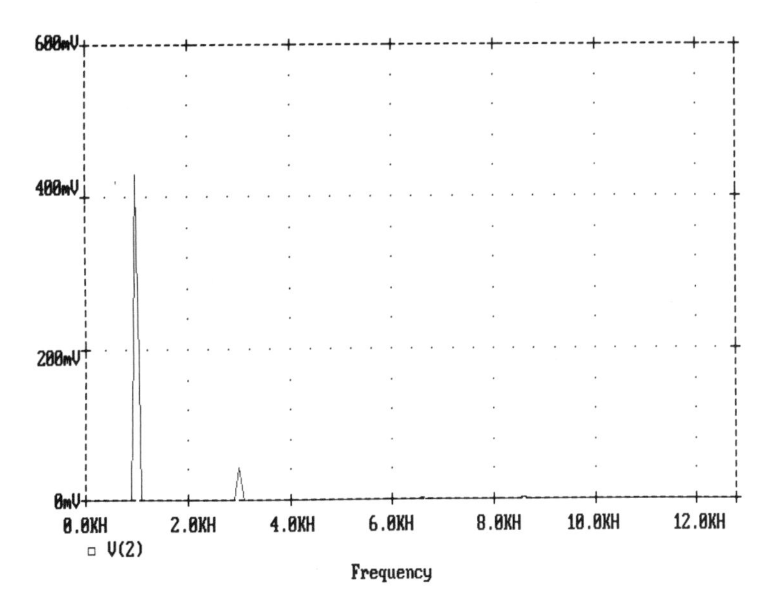


**Figure 10.5** Plot of output signal (spectral domain).

The primary source of error in the transform comes from the interpolation done by Probe to obtain evenly spaced samples, that is, the time points in the transient run did not occur at the right places (this need not be true of transforms of a frequency analysis, where you can control the analysis interval). Using interpolated values is fine so long as you realize that this ''smooths'' the results, which is a way of saying that the signal is being processed, **before being transformed,** by a low-pass filter. This filter's transfer ratio is unity at DC, zero at the Nyquist critical frequency, and some value in between. This is why the results we obtain for harmonic magnitude using the FFT in Probe are always lower than the results from the .FOUR statement. The effective bandpass of this ''smoothing'' filter is increased by using more samples in the FFT, which means forcing PSpice to take more time-steps.

**Exercise 10.6.1**

Improve the accuracy of the measurement of the third harmonic by doubling the number of steps in the Fourier transform. You will do this by decreasing the maximum step size of the transient simulation. Notice that the X-axis extends to twice the range as our example. How does the new ''spike'' at 300 hertz compare with the value from the harmonic decomposition?

## 10.7 INTERMODULATION DISTORTION

Having just tried an example of calculating spectral values, let's broaden our scope to calculate intermodulation distortion. ''Intermodulation'' is the name for the process by which any signal processor, such as an amplifer, converts superimposed signals into signals whose frequencies are the sum and difference of the frequencies of the input signals. The nonlinearities which cause this to happen also produce distortion products at other frequencies, such as twice the frequency of signal A, minus the frequency of signal B, and other such combinations. Since most of these combinations are not harmonically related to the input signals, these distortion products are considered to be the most objectional in, say, a high-fidelity audio system. There are many ways, or standards, for expressing the amount of distortion of this type, however we shall look at a total calculation as a guide to any method that you might prefer.

Using a variant of our previous example, we inject two signals into the polynomial function so that the peak input barely saturates the function. Using the frequencies of 800 and 1000 hertz to present one style of measuring intermodulation distortion, where the input frequencies are close together, the circuit file becomes:

```
*intermodulation distortion
.opt itl5=0
i1 0 1 sin(0 .28 1000)
i2 0 1 sin(0 .28 800)
```

```
rin 1 0 1
e3  2 0 poly(1) (1,0) 0 1 0 -1
r3  2 0 1G
.tran 50u 50m 0 50u
.probe
.end
```

After running PSpice, Probe is again used to transform the results into the signal spectra. We can display the input signal to check that the simulation and transform went well (see Fig. 10.6). Now, we transform the waveform to show the input spectra (see Fig. 10.7). At this time, you might want to check your calculating skills by analyzing the input spectrum, which is known. The peaks at 800 and 1000 hertz are single frequencies, so we divide by the number of bins in the subcritical frequency band to get the magnitude of each component:

0.2788 at 800 hertz
0.2780 at 1000 hertz

which are very close to the 0.28 input magnitude. Notice that the higher frequency signal's magnitude is lower, a manifestation of the "smoothing" filter caused by using interpolated values. Now, looking at the output spectra, in Figure 10.8, we see many more spectral lines. Most of these are intermodulation distortion products. Let's start to calculate.

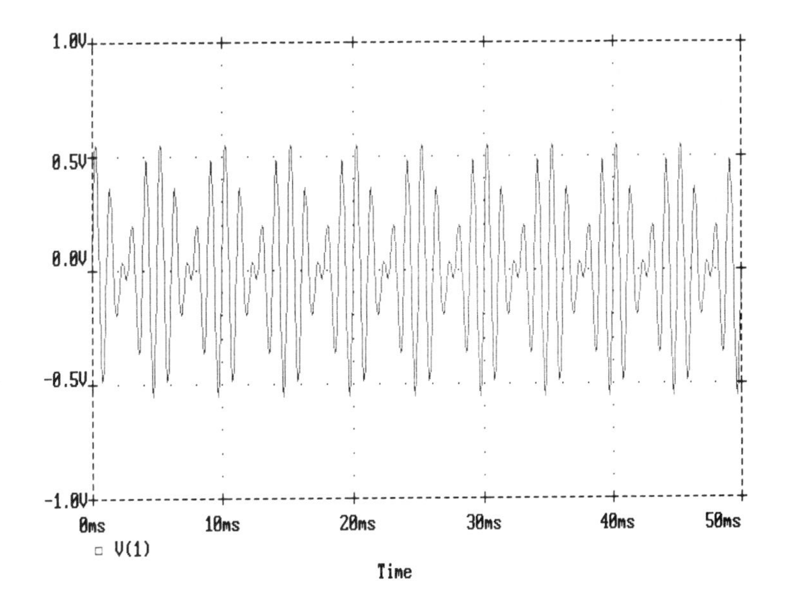


**Figure 10.6** Plot of input signal (time domain).

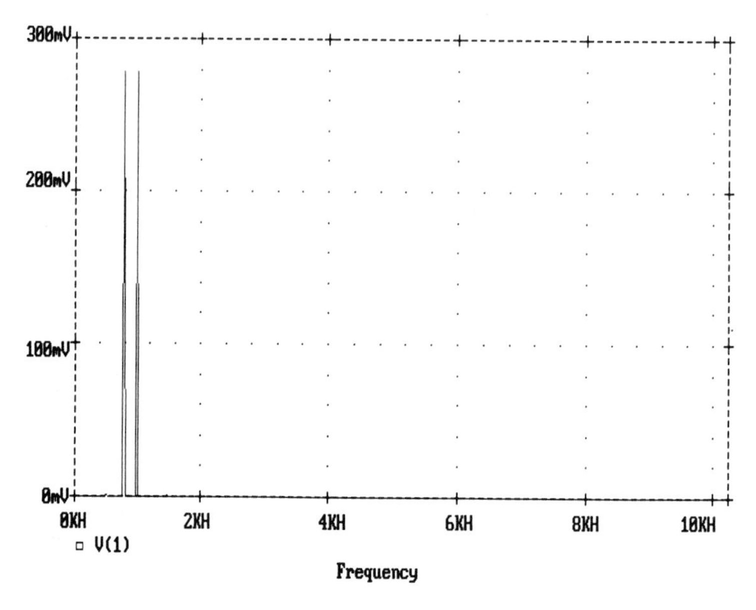


**Figure 10.7** Plot of input signal (spectral domain).

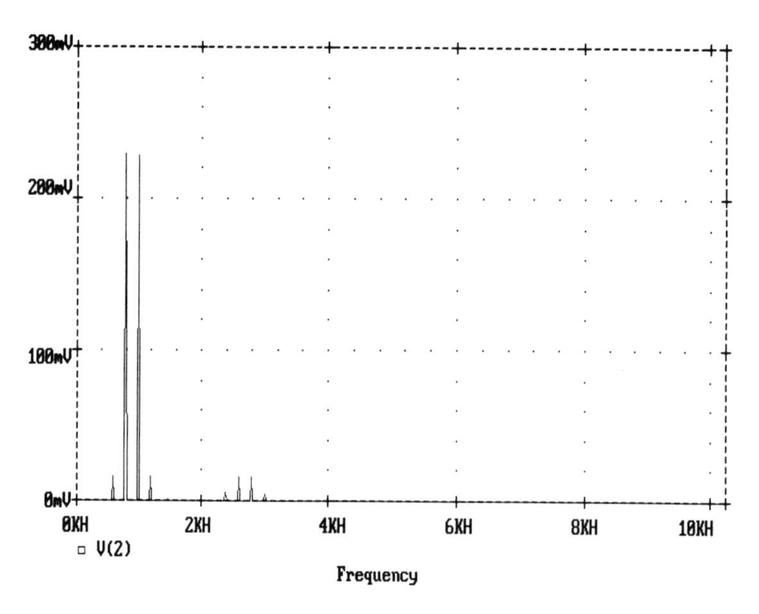


**Figure 10.8** Plot of output signal (spectral domain).

First, we calculate the magnitude of the fundamental outputs:

0.2296 at 800 hertz
0.2298 at 1000 hertz

Then, there is harmonic distortion, most of which is at the third harmonic of the input:

5.240E−3 at 2400 hertz
5.102E−3 at 3000 hertz

These combine to yield harmonic distortion of

$$(0.005240^2 + 0.005102^2)^{1/2}/(0.2295^2 + 0.2287^2)^{1/2} = 2.25\%$$

which looks good. Even though the signal reaches the clipping level, the individual distortion energies do not amount to much.

Finally, there is intermodulation distortion:

16.43E−3 at 600 hertz
16.29E−3 at 1200 hertz
15.58E−3 at 2600 hertz
15.54E−3 at 2800 hertz

These combine in the same root-sum-square fashion to yield

$$(0.01643^2 + 0.01629^2 + 0.1558^2 + 0.01554^2)^{1/2}/(0.2295^2 + 0.2287^2)^{1/2} = 9.83\%$$

which is about four times the combined distortion energy of the harmonic distortion.

# CHAPTER 11

# Device Models

Device models (or just "models") are SPICE's way of collecting the operating characteristics of a circuit element ("device"). So far we have worked only with fairly simple devices, such as resistors and capacitors, and even these devices can make good use of having a model. But especially for active devices, such as diodes and transistors, it is essential to collect the numerous parameters that describe how the device will behave and then refer to that particular set of parametric values by a shorthand name. This lets you label each instance of the device, in the circuit file, by a name that is convenient and mnemonic. Furthermore, when you decide to change the model parameters, this needs to be done only in the device model and it will affect all of the devices in your circuit file which refer to that model.

## 11.1 THE .MODEL STATEMENT

The .MODEL statement sets aside a set of parametric values for reference by devices in PSpice. Not every device needs a model; for example, resistors that do not refer to a model are assumed to have a constant resistance value for all simulations. Every device which does refer to a model must have that model defined, which means it needs to have a .MODEL statement that completes the description for how the device will operate. The syntax for the statement is

.MODEL <*name*> <*type name*> ([<*parameter name*> = <*value*>]*)

The first <*name*> is the shorthand label or "model name" that you want to refer to the device as. Often this is a manufacturer's part number, such as "2N3904" for a transistor, or a descriptive name, such as "FILM" for a metal film resistor. You may use any name that conforms to the naming conventions of the simulator; they must begin with an alphabetic character, and continue with alphabetic or numeric characters, or "_" and "$". For example, 2N3904 is usually modified to be Q2N3904 to fit the naming conventions.

The *<type name>* is a device "type" description, which may be one of the following for the linear devices:

| | |
|---|---|
| CAP | for a capacitor |
| IND | for an inductor |
| RES | for a resistor |

one of the following for the semiconductor devices:

| | |
|---|---|
| D | diode |
| NPN | bipolar transistor (NPN) |
| PNP | bipolar transistor (PNP) |
| NJF | junction field-effect transistor (N-channel) |
| PJF | junction field-effect transistor (P-channel) |
| NMOS | MOS field-effect transistor (P-channel) |
| PMOS | MOS filed-effect transistor (P-channel) |
| GASFET | gallium-arsenide field-effect transistor (N-channel) |

and the following for the "miscellaneous" device group:

| | |
|---|---|
| CORE | nonlinear magnetics |
| VSWITCH | voltage-controlled switch |
| ISWITCH | current-controlled switch |

Any of the parameters allowed for the device model are thus defined. If you do not include a model parameter and value, then there is a default value which will be used instead. Usually these default values are set to a convenient value which give a typical operation, or are set so they have no effect on the device's operation (which means you may ignore them if they are of no interest). Let's look at what these parameters can do for your circuits.

## 11.2 MODELS FOR PASSIVE DEVICES

When SPICE users talk about "models," they are usually referring to models for semiconductor devices. This is an important area; design groups that use circuit simulators often maintain libraries of models for their work, and may even have some engineers whose only job is to develop new model sets. However, even simple devices may have models. For example, the model for the capacitor includes parameters for the following:

C which is the multiplying (scaling) factor, default value = 1
VC1 which is the linear voltage coefficient, default value = 0
VC2 which is the quadratic voltage coefficient, default value = 0

and others, which we will look at later. Already you may suspect that we will be able to define a type of capacitor whose capacitance will vary with the voltage impressed on the plates by setting the voltage coefficients.

To see how to use the device models, let's recall the syntax for including a capacitor in a circuit file, which is

C<*name*> <*node*> <*node*> [*model name*] <*value*>

This is different from how you are used to writing it since it includes the [*model name*], which is an optional item (we have not exercised the option until now). So, you can have a circuit with capacitors that may include a reference to a model, as in this fragment of a circuit file:

```
C5 2 7 .015
C6 3 5 mod_cx 2
C7 4 6 mod_cx 1
.model mod_cx cap(vc1 = .1 c = .001)
```

In this case the value of C5 is 0.015 farads, but the value for C6 and C7 depends on the model parameters. Their values are calculated by the formula

$$\text{capacitance} = \langle value \rangle \cdot \text{C} \cdot (1 + \text{VC1} \cdot \text{voltage} + \text{VC2} \cdot \text{voltage}^2)$$

This means that, with no voltage across the capacitors, the value for C6 is 0.002 farad and the value for C7 is 0.001 farad. As the voltage varies, C6 and C7 will change their values by 10 percent per volt, where the voltage is the difference of the first node from the second node (so it matters, now, which way you connect the capacitor).

## 11.3 SCALING COMPONENT VALUES

Using the previous example circuit fragment, since C6 and C7 both refer to the "mod_cx" capacitor model, you may scale their values relative to each other; that is, regardless of the value calculated by the formula above, the value of C6 **will always be twice** the value of C7 (not strictly true, since this example shows a capacitor model which is also voltage dependent). You may shift the values of a whole set of capacitors in your circuit without changing their relative values, by changing the model to, say

```
.model mod_cx cap(vc=.1 c=.002)
```

to double their zero-bias values to 0.004 and 0.002 farads, respectively.

Another way to set up capacitors that scale is to start with the actual component values you would normally use; for example:

```
C5 2 7 .015
C6 3 5 mod_cx .002
C7 4 6 mod_cx .001
.model mod_cx cap(c=1)
```

so that the values for C6 and C7 are still 0.002 and 0.001, respectively, because the multiplier parameter is now 1. Then, should you decide later to increase the values for the "cmod" capacitors by 10 percent, you would change the model to

```
.model mod_cx cap(c=1.1)
```

You may use as many models as you like in a circuit file, although each component may refer to only one model. If you are crafty, you may find ways to set up "generic" circuit blocks, such as filter modules, which are customized by setting the value of the multiplier parameter.

As you might have suspected, the resistors and inductors in PSpice may have models also. Just like the "C" multiplier:

R is the resistance multiplier parameter in the RES model
L is the inductance multiplier parameter in the IND model

This means we could have used scaling models for all of the elements in our LC-filter example, as shown for one of the filter sections:

```
*Q=1
R2 1 4 rmod 1
L2 4 5 lmod 1
C2 5 0 cmod 1
.model rmod res(r=100)
.model lmod ind(l=10m)
.model cmod cap(c=1u)
```

If we did this correctly for all of the filter sections, we could shift the natural frequency of the filters while keeping the same "Q" values.

## 11.4 SWEEPING COMPONENT VALUES

In PSpice, the .DC statement has been generalized to sweep model parameters which in turn will sweep the component value, as well as (normally) sweeping voltages or currents. This is done by extending the meaning of <*source name*> to include references to model parameters; for example:

```
.DC vin 2 12 2
```

sweeps the value of the voltage source "vin" from 2 volts, to 12 volts, in 2-volt increments. By example, we could also sweep the resistance multiplier in "rmod" with the following circuit file fragment:

```
R7 4 6 rmod 1
.model rmod res(r = 100)
.DC res rmod(r) 100 150 10
```

which will sweep the "R" parameter of the "RES" model named "RMOD," starting with R = 100, in increments of 10, until R = 150. This will change the multiplier for R7 and any other resistors that reference this model.

## 11.5 TEMPERATURE ANALYSIS

Another common way to check a circuit is to operate it at different temperatures to verify that certain performance standards are met. SPICE and PSpice both include a control statement of the form

.TEMP <*value*> . . .

with a list of values specifying the temperatures, in degrees Centigrade, at which all of the other analyses (such as DC, AC, etc.) are to be run; that is, the .TEMP statement acts as an "outer loop" for all of the other analyses. When the temperature is changed, PSpice recalculates internal values using the new temperature (e.g. the noise contribution for resistors involves the factor $4 \cdot k \cdot T$) and also makes adjustments where the device models have a temperature dependence. For example, the device models of the capacitor, inductor, and resistor all have parameters for temperature dependence, which are

TC1 a linear dependence on the change in temperature, in % per °C
TC2 a quadratic dependence on the change in temperature, in % per °C$^2$

so that, just looking at the temperature-related factors, the formula for the value of a resistor with a model is

$$\text{resistance} = \langle value \rangle \cdot R \cdot (1 + TC1 \cdot (T - T_0) + TC2 \cdot (T - T_0)^2)$$

where $T_0$ is nominally 27°C (which may also be set, using the .OPTIONS statement, though this is rarely done). The capacitor and inductor follow the same form. Other devices, such as transistors, have temperature dependencies built into the (more complicated) physical models for these devices, so that the user does not have to include the factors directly for operation at different temperatures.

## 11.6 SWEEPING TEMPERATURE

Very much the way PSpice is able to sweep component values, you may also have PSpice sweep the value for temperature. The calculations that are done are the same as those for the .TEMP statement, however the simulation is done as a DC sweep so that graphical results are available to Probe. Moreover, since you are allowed to nest a DC sweep within another, you will be able to sweep a source or component value while stepping the temperature (or vice versa)!

The form of the DC sweep statement to sweeping temperature is similar to what we have used for sweeping component values; for example:

```
.DC TEMP 30 50 5
```

will sweep temperature, called "TEMP" by PSpice, from 30°C to 50°C in 5°C increments.

# CHAPTER 12

# Active Devices

Active devices, such as diodes and transistors, are at the heart of why SPICE was developed and became so successful. The behavior of these devices constrains the mathematics and numerics of the simulator, and only a few of the algorithms that could be used have proven successful in the face of the large changes in conductance that active devices undergo in normal circuit operation.

In this chapter we will take a qualitative tour of the models in PSpice (which are compatible with the devices in SPICE2.G6). The tables of parameters (with their names, description, units, and default values) **will not** be in this chapter, as your SPICE manual will have these (there are some in Appendix B of this book). Neither will you find the full set of equations for each device, which tend to obscure the external operation with which the user is familiar. Instead we will focus on the terminal characteristics that electrical engineers know and show how the equations, with their copious parameters, describe and model these characteristics.

## 12.1 ACTIVE DEVICE MODELS

The semiconductor diode is usually the first active device students learn about; its ability to change resistance and switch, depending on current direction, is the basis for most beginning electronics courses. Those courses, including semiconductor physics, will trace the development of the celebrated Shockley equation controlling *pn* junction current, which is

$$\text{junction current} = I_s \cdot (e^{V/(k \cdot T/q)} - 1)$$

It is sets of equations, like the Shockley equation, that define the operation of active devices for SPICE. One of the options for the PSpice program is the source code for the routines containing these equations (albeit in a form that is efficient and suitable for the simulator) should you want to try different physics.

The parameters that are available through the .MODEL statement are ones that appear in the controlling equations for the device. For example, in the Shockley equation, the parameter "IS" may be specified in the diode model. Of course, $k$ and $q$ are physical constants, and "T" is specified as the temperature for the simulation. In this way the user controls the device operation without writing new equations for each device.

Models for active devices are similar to those for passive devices, just more complicated in the conductances and currents that are calculated. For SPICE, it is not enough to consider, say, forward current gain for the bipolar transistor as an isolated feature of that device. All of the operating characteristics must be combined into a unified model, since SPICE is not capable of knowing when to discard effects that, for the circuit condition at hand, are negligible (which, of course, is a time-honored engineering practice). All of the characteristics that affect the calculation of conductance, transconductance, current, et cetera, must be present each time the device is evaluated. This means that device operation, which we normally split into operating "regions" (for example, "saturation" or "cutoff"), becomes one continuous set of formulas. It is difficult to develop device models that behave this way.

The benefit, for the SPICE user, is that all of the device characteristics can be included in the simulation (of course, you may choose to ignore some characteristics). Often you find that a circuit that you expect to work does not simulate the way you expect because it has "been had" by some device characteristic that you overlooked during design. This is the purpose of SPICE: to verify the operation of a circuit (after all, the simulator is dumb and will not be misled by what you intended the circuit to do). This is why you want models that are complete enough to not only simulate your circuits when they behave as you expect, but also to show you when they don't. So, the models are important.

There is only one model, in SPICE, for each device type. This model is the full nonlinear set of equations describing currents, conductances, and capacitances. New SPICE users often wonder if, for AC analysis, the simulator uses a hybrid-pi model for the bipolar transistor. It does, but it is imbedded in the nonlinear equations, which are used to arrive at the bias-point for the circuit and then the conductance, transconductance, and capacitance values are saved for use by the AC analysis section of the simulator. Think of it as PSpice calculating a hybrid-pi model for each transistor in the circuit. But the topology internal to the transistor is the same for transient and AC analysis. For the latter analysis, the small-signal values are calculated and used.

## 12.2 SEMICONDUCTOR DIODE

The diode model in PSpice, as mentioned before, contains a nonlinear current source which follows the Shockley equation:

$$Id = Is \cdot (e^{Vd/(n \cdot Vt)} - 1)$$

where

Vd is the voltage across the junction

Vt is the thermal voltage ($k \cdot T/q$)

to model the current-voltage effects of the semiconductor junction. This does not include the nonideal operation of real diodes. For example, at low currents (less than 1nA), other semiconductor processes that increase the flow of currents become noticeable. As a practical matter, these are small currents which are ignored by SPICE.

As you can see in Figure 12.1, by setting IS to different values you can obtain the characteristics of (i) a Schottky-barrier diode, and (ii) a silicon diffused-junction diode. High-current effects are modeled, grossly, by including a series resistance which is intended to combine the effects of bulk resistance (the material on each side of the junction) and high-level injection. At high currents the observed diode current stops following the Shockley form

$$Id = Is \cdot e^{Vd/(n \cdot Vt)}$$

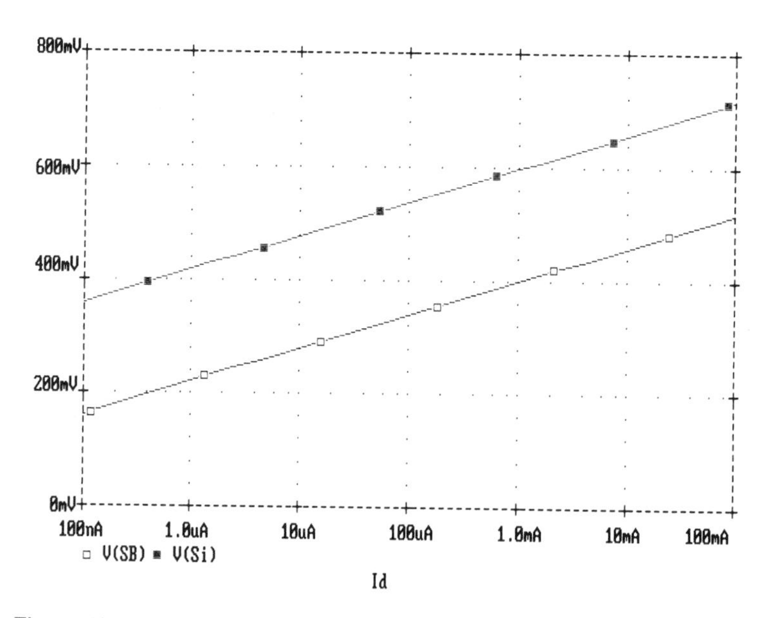


**Figure 12.1** Plot of device current versus voltage using IS values typical for Schottky-barrier and diffused-junction diodes.

and approaches a modified form

$$\text{Id} = \text{Is} \cdot e^{\text{Vd}/(2 \cdot \text{n} \cdot \text{Vt})}$$

Again, for practical reasons (SPICE's emphasis is on integrated circuits, which rarely develop such large currents in normal operation), the simulator does not include this modified form. Instead, SPICE has only the series resistance parameter, "RS," available for more limited modeling of this effect. It is not so important to model the effect accurately as to make a provision for the effect so it is available to indicate abnormal operation.

As you can see in Figure 12.2, by setting RS you can limit the exponential effect of the Shockley equation and the device becomes resistive.

For reverse operation, the value of "Is," which the Shockley equation asymptotically approaches, is usually too small a value as real devices have leakages which allow current across the junction. To help model this, as well as improve the operation of the simulator, a minimal conductance is connected in parallel with the junction. The value of the conductance is set by the GMIN option (see your user manual for details of the ".OPTION" statement).

As you can see in Figure 12.3, the reverse diode current deviates from the Shockley equation due to the GMIN conductance in parallel with the junction. Reverse breakdown, as found in Zener diodes, is modeled by another exponential form

$$\text{breakdown current} = \text{Ibv} \cdot e^{-(\text{Vd} + \text{Bv})/\text{Vt}}$$

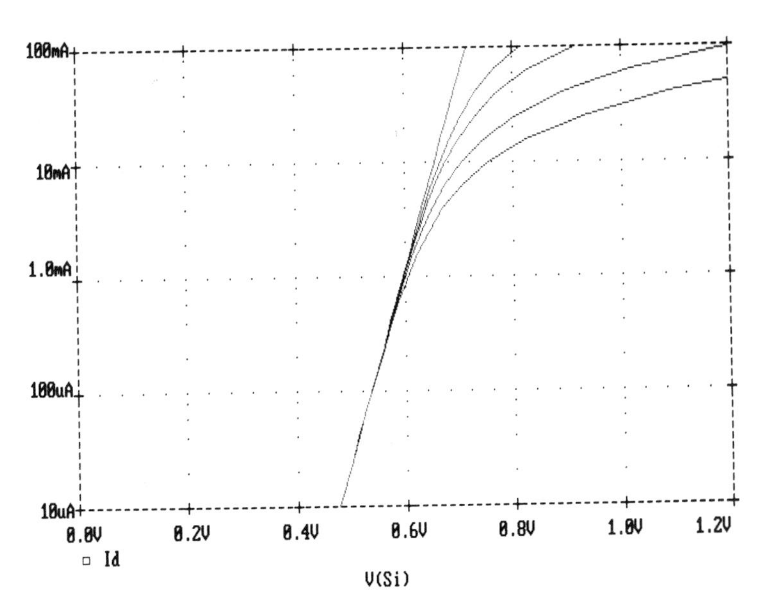


**Figure 12.2** Plot of device current versus voltage for different values of RS.

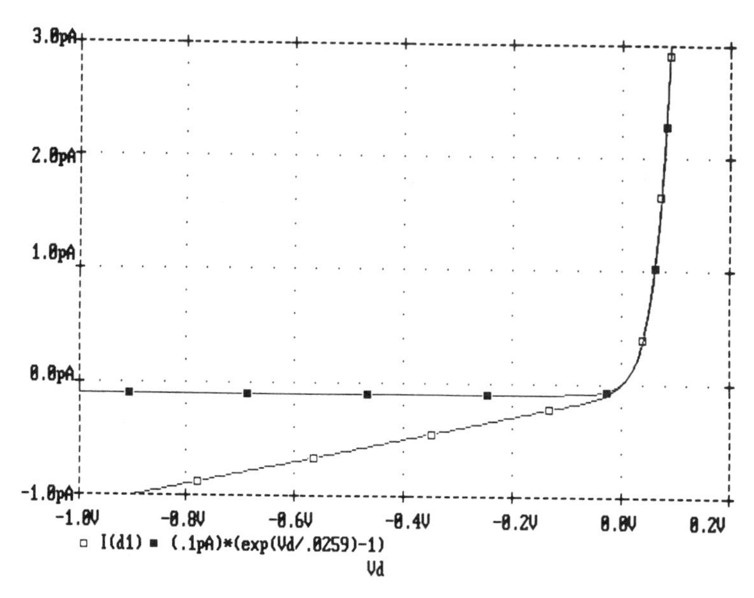


**Figure 12.3** Plot of device reverse current versus voltage and Schockley equation.

which gives the correct, but not exact, effect of reverse breakdown. One problem is that semiconductor junctions have more than one breakdown mechanism, and these processes can occur at the same time. Again, as a practical matter, SPICE does not attempt to model the "blend" of processes; the simple form serves most engineering purposes. Any particular breakdown curve could as easily be modeled over a wide range using a controlled-current source defined by the user.

Figure 12.4 shows both forward and reverse operation. Note that this figure uses extraordinary parameter values to exaggerate the differences from the Shockley equation. Diode capacitance is modeled by a voltage-dependent capacitor, which is connected in parallel with the nonlinear current generator described previously, to represent the charge storage effects of the junction. There are two components to this charge: (i) the reverse-voltage capacitive effect of the depletion region, and (ii) the forward-voltage charge represented by mobile carriers in the diode junction.

Reverse-voltage capacitance follows the simple approximation that the depletion region (the area of the junction that is depleted of carriers) serves as the gap between the "plates" of a capacitor. This region varies in thickness, and therefore the capacitance varies with applied voltage. For a step (abrupt) junction, or linearly graded junction, the capacitance approximation is

$$\text{capacitance} = \text{Cjo}/(1 - \text{Vj/phi})^{\text{M}}$$

where Cjo is the zero-bias value, "phi" is the junction barrier potential, and "M" is the grading coefficient which varies (½ is used for step junctions and ⅓ is used for linearly graded junctions, and most junctions are assumed to be somewhere in between).

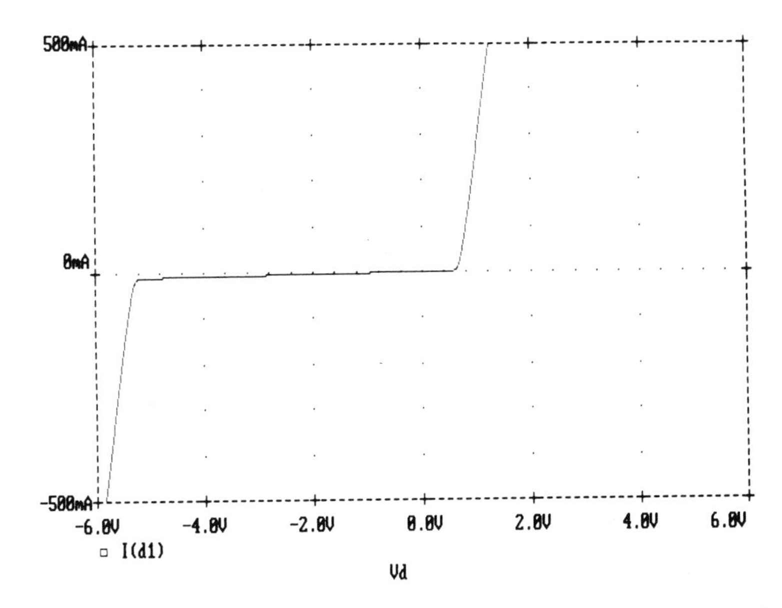


**Figure 12.4** Plot of "full range" device current versus voltage (using exaggerated values).

There is often confusion about the barrier potential, "phi," which appears in the capacitance equation. From capacitance measurements "phi" takes on a value of nearly 0.7 volts for regular (silicon) junction diodes, and a range of 0.58 to 0.85 volts for various Schottky-barrier diodes. This value sometime gets confused with the forward-current voltage drop of the diode, and sometimes gets confused with the energy gap of the material.

As you can see in Figure 12.5, varying M will generate a variety of reverse-bias capacitance characteristics. Some inspection of the capacitance formula reveals that it predicts infinite capacitance for a forward bias, which is not the case for a real junction. Several depletion capacitance formulas have been proposed which more correctly fit observed operation, however SPICE uses a simple approach: for forward biases beyond some fraction (set by the parameter FC) of the value for "phi," the capacitance is calculated as the linear extrapolation of the capacitance at the departure. This provides a continuous numerical result, and does not affect circuit operation significantly because for forward bias the device capacitance is normally dominated by diffusion capacitance.

The diffusion charge varies (and therefore capacitance) with forward current and is simply modeled as a "transit time" for the carriers to cross the diffusion region of the junction. The total charge is

$$\text{diffusion charge} = \text{device current} \cdot \text{transit time}$$

and capacitance is the derivative, with respect to bias, of this

$$\text{diffusion capacitance} = \text{TT} \cdot \text{Is}/(\text{n} \cdot \text{Vt}) \cdot e^{\text{Vd}/(\text{n} \cdot \text{Vt})}$$

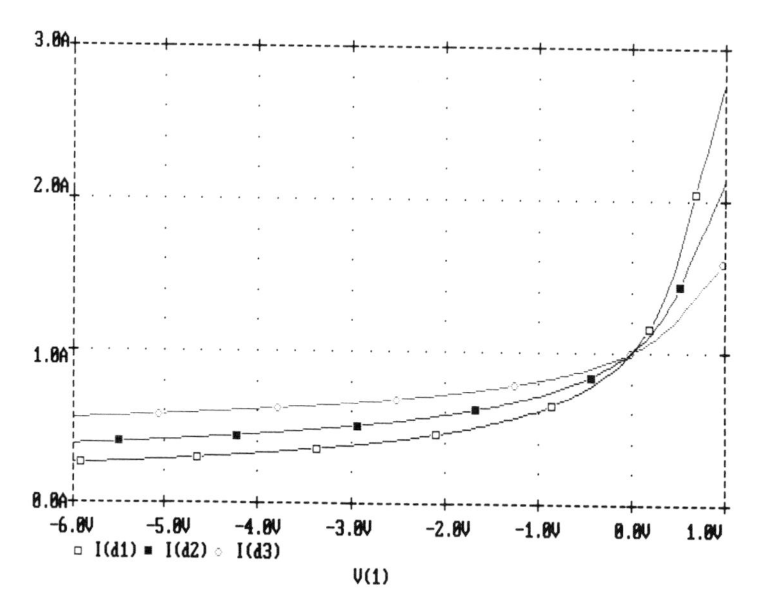


**Figure 12.5** Plot of junction capacitance versus voltage.

Difffusion charge manifests itself as the ''storage time'' of a switching diode, which is the time required to discharge the diffusion charge in the junction, which must be done before the junction can be reverse-biased (switched off). Storage time is normally specified as the time to discharge the junction so that it is supporting only a fraction (typically 10 percent) of the initial reverse current. First, a forward current is supplied to the device to charge the junction. Then, as quickly as possible, a reverse current is supplied to the device. Internally, the junction is still forward-biased to a voltage nearly the same as before the switch in current; the junction is still conducting at the forward current rate. This internal current adds to the external current as the total current discharging the junction. As the junction voltage decreases, the internal current falls off exponentially (according to the Shockley equation). The system is a relatively simple differential equation which can be solved to an explicit equation for the TT parameter (assuming complete discharge) as follows:

$$\text{transit time} = \text{storage time}/ln((i_F - i_R)/-i_R)$$

As you can see from Figure 12.6, the diffusion charge dominates the reverse recovery characteristic of the diode. During the last part of the recovery, as the junction becomes reverse-biased, the depletion capacitance dominates. This causes the small tail at the end of the discharge cycle. Total capacitance is taken to be the sum of these capacitances: the depletion approximation dominates for reverse bias as the device current is small, and the diffusion approximation dominates for forward bias as the device current is large.

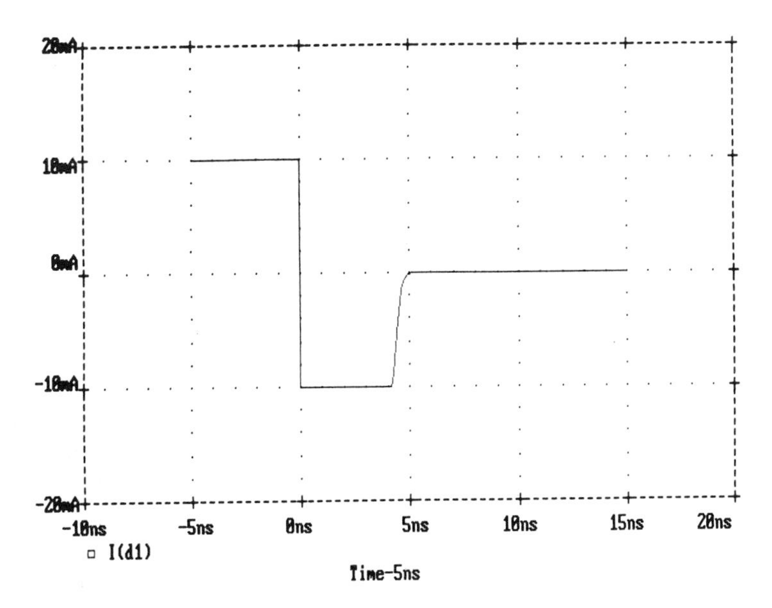


**Figure 12.6** Plot of reverse recovery current (transient analysis).

## 12.3 JUNCTION FIELD EFFECT TRANSISTOR (JFET)

The JFET is the simplest of the transistor devices. In this device the increase in the depletion region by gate junction bias "pinches" the channel, increasing its resistance to drain current. It is known as a "square law" device because of the expression relating drain current to gate junction voltage:

$$\text{Id} = \text{beta} \cdot (\text{Vgs} - \text{Vto})^2$$

While actually an approximation of the transfer function given by the exact analysis of channel charge, this is almost universally used (see Fig. 12.7). Another way of arriving at the same square law relation is by making the approximation that the gate junction capacitance is a linear function of the gate junction voltage (which in turn describes how the channel region is modulated). As we saw for the diode, reverse-bias capacitance is not a linear function but it may be approximated that way for biases much larger than the barrier potential ("phi") of the junction. The error associated with using the square law form happens to be quite small (when compared to exact analysis as well as real devices). The square law result applies only when Vds is greater than Vgs-Vto (the "pinch-off" voltage), when the channel of the FET is "saturated." When Vds is below pinch-off, the expression relating drain current to gate junction voltage is

$$\text{Id} = \text{beta} \cdot (2 \cdot (\text{Vgs-Vto}) \cdot \text{Vds} - \text{Vds}^2)$$

which describes (on an Id versus Vds plot) an inverted parabolic curve passing through the origin and which, at its peak value (when Vds is at pinch-off), intersects

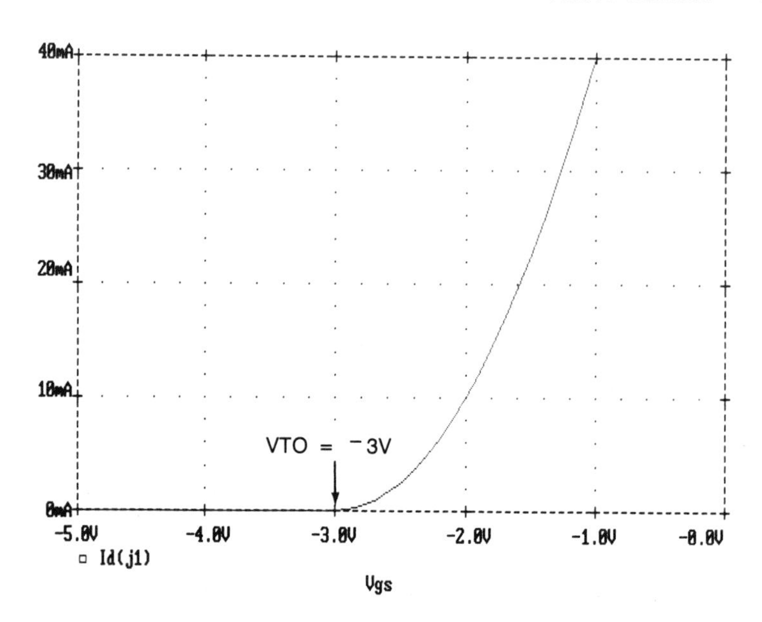


**Figure 12.7** Plot of drain current versus gate-source voltage.

the square law formula. This parabolic region of operation is called the "linear" region; for small drain voltages, the expansion of the equation (above) is dominated by the linear term

$$\text{Id} \approx 2 \cdot \text{beta} \cdot (\text{Vgs-Vto}) \cdot \text{Vds}$$

Finally, Ids is zero when Vgs is less than Vto (see Fig. 12.8).

Real JFETs, in the saturation region, are not ideal current devices since their drain currents vary with drain voltage. This effect is modeled by the device parameter LAMBDA, which sets the output conductance

$$\text{Id} = \text{BETA} \cdot (1 + \text{LAMBDA} \cdot \text{Vds}) \cdot (\text{Vgs-VTO})^2$$

which yields an increasing current for increasing values of Vds, as shown in Figure 12.9.

Since transconductance is

$$d\text{Id}/d\text{Vgs}$$

we can readily relate BETA to transconductance by differentiating the drain current formula to get

$$\text{transconductance} = 2 \cdot \text{BETA} \cdot (1 + \text{LAMBDA} \cdot \text{Vds}) \cdot (\text{Vgs-VTO})$$

The capacitances of the JFET follow the form we saw for the diode. Both the gate-to-source and gate-to-drain junctions have a nonlinear capacitor. The zero-bias capacitance value is selected for each junction. When these junctions become

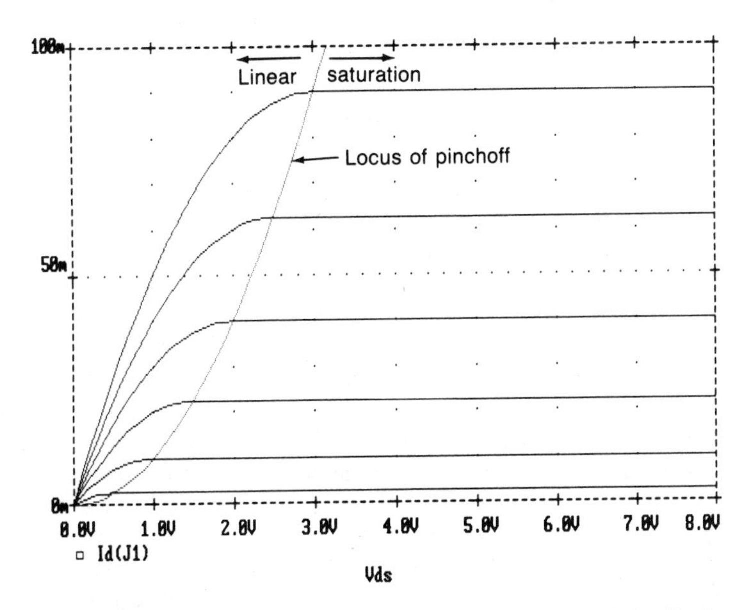


**Figure 12.8** Plot of drain current "curve family," with locus of pinch-off voltage.

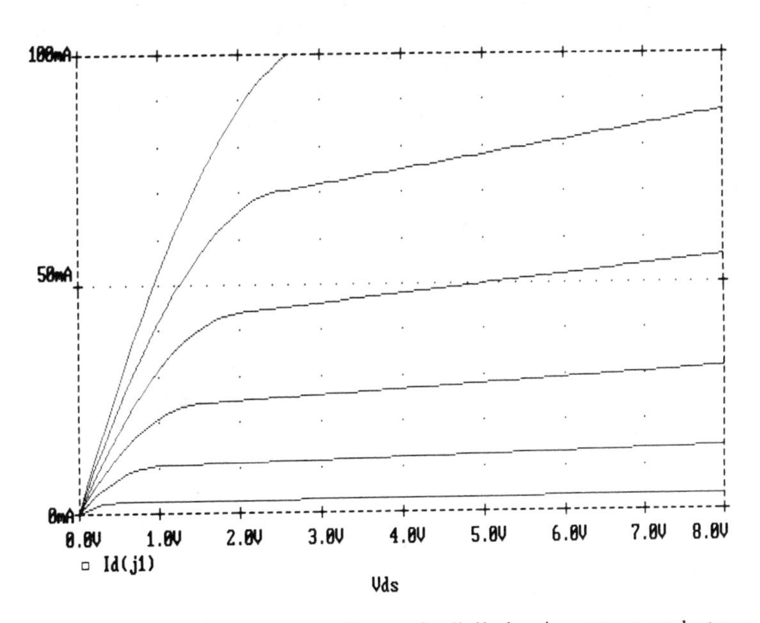


**Figure 12.9** Plot of drain current "curve family" showing output conductance in saturation region.

forward-biased, the straight-line extension of capacitance is used (just like the diode). However, there is no provision for diffusion charge in the junction since JFETs are rarely used in a mode that has either junction forward-biased.

## 12.4 GALLIUM-ARSENIDE MESFET (GaAsFET)

The GaAsFET is a Schottky-barrier gate FET, or MESFET (for *ME*tal *S*emiconductor FET), made of gallium arsenide (the "GaAs" comes from the chemical abbreviation for the material). The primary advantage of GaAs over silicon is its electron mobility, which is six times greater (mobility is the speed of electrons in the material for a given electric field, which propels the electrons). This is an advantage of GaAs that is important for high-frequency electronics. At present the major drawback to GaAs is the difficulty in processing and manufacturing devices, however these problems are being solved quickly due to a large market for high-speed devices in computing and defense electronics. Soon the fastest computers, using GaAs devices in modules designed using PSpice simulations, will be simulating tomorrow's circuits.

The GaAs MESFET operation is like the silicon MOSFET. An insulating layer between the gate and channel is provided by the potential barrier formed at the contact of two materials, in this case a metal gate and GaAs substrate. Similar to the MOSFET a channel charge is induced to create a conducting path under the gate, connecting the drain and source of the device. However, the detailed device operation is different in that in GaAs the electron velocity "saturates" for an electric field roughly ten times lower than in silicon. Thus the saturation in drain current, for GaAs, occurs due to carrier-velocity saturation, whereas channel pinch-off causes this in silicon. There are several proposed models for the conductivity of the channel, so PSpice includes both the Curtice model and the Raytheon model.

The Curtice model was one of the first to be implemented in a circuit simulator, circa 1983, and therefore gained early acceptance. The formula for active operation, where Vgs is greater than Vto, is

$$\mathrm{Id} = \beta \cdot (\mathrm{Vgs\text{-}Vto})^2 \cdot (1 + \lambda \cdot \mathrm{Vds}) \cdot \mathit{tanh}(\alpha \cdot \mathrm{Vds})$$

which includes both "linear" and "saturated" operation and is an empirical fit using the hyperbolic tangent function. The device capacitances for the Curtice model are simple: the normal *pn* junction capacitance is used for the gate-source and gate-drain capacitance, and a fixed capacitance is available for drain-source capacitance modeling.

The Raytheon model (named after the employer of the developers) is a more recent model, circa 1986, and benefits from the later research into GaAs devices. In particular, it has two improvements over the Curtice model: (i) an enhanced drain current formulation and (ii) a new capacitance model. The drain current formula was modified; while the Curtice model did well for the observed change in drain

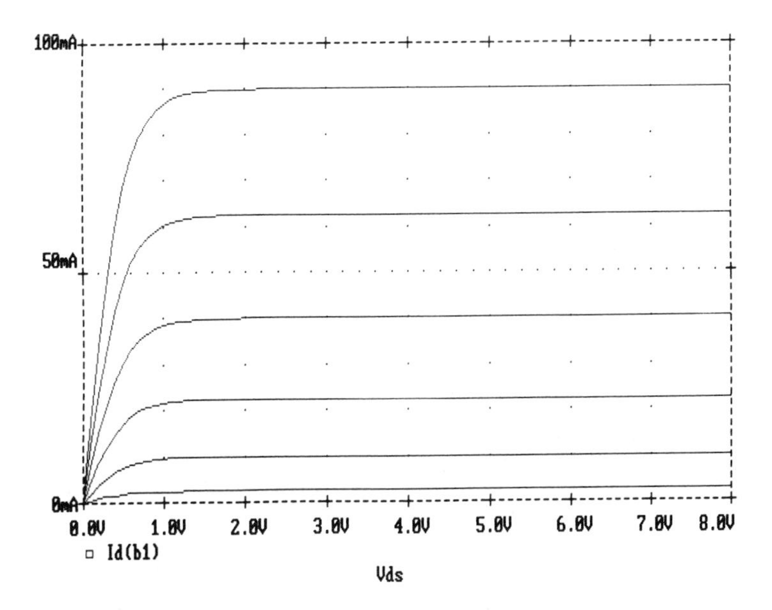


**Figure 12.10** Plot of drain current "curve family" for Curtice model.

current versus Vds, it did not do well for drain current versus Vgs. Using observed operation, the portion of the Curtice formula

$$Id = \beta \cdot (Vgs\text{-}Vto)^2$$

was changed to

$$Id = \beta \cdot (Vgs\text{-}Vto)^2/(1 + b \cdot (Vgs\text{-}Vto))$$

to effect the desired operation for Vgs >> Vto. To increase computational efficiency, the hyperbolic tangent function was replaced with a polynomial approximation

$$tanh(x) \approx 1 - (1 - x/3)^3$$

and when x is larger than 3 unity is used (this defines the onset of device saturation). Using Figure 12.11, compare the Raytheon model to the Curtice model.

The new capacitance model comes (again) from the problem of carrier-velocity saturation. In theory, the gate-source capacitance should increase abruptly at the onset of velocity saturation and the gate-drain capacitance should decrease abruptly. In practice, velocity saturation occurs more gradually so the capacitance changes will not be as abrupt. The Raytheon capacitance model is "charge oriented" to calculate the effects of velocity saturation. Once the channel charge and therefore capacitance (as the change in charge versus voltage) is calculated, it is split into gate-source and gate-drain capacitance values. Furthermore, these capacitance values maintain symmetry if the device is operated in the inverted mode.

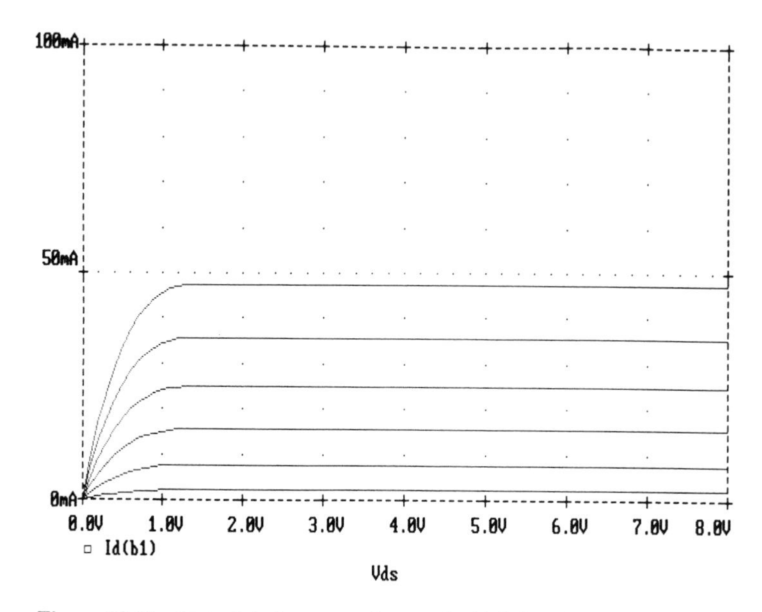


**Figure 12.11** Plot of drain current "curve family" for the Raytheon model.

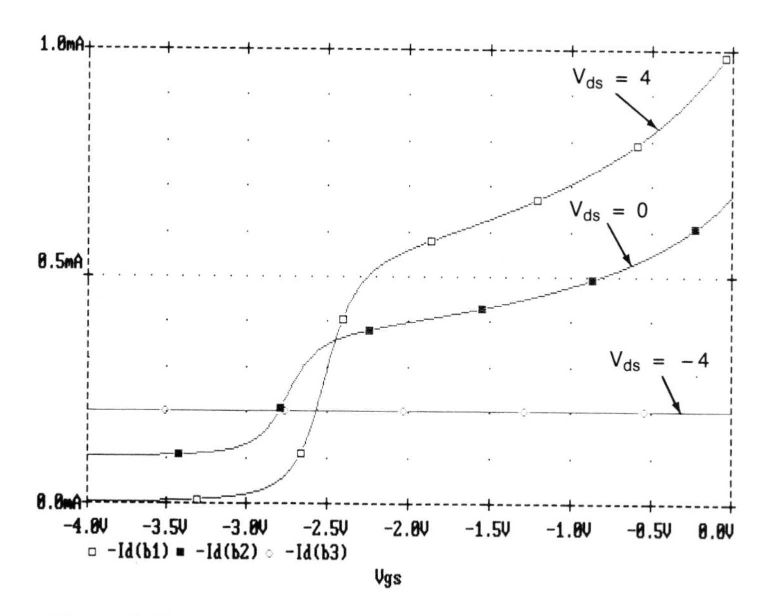


**Figure 12.12** Plot of Raytheon device capacitances versus Vds and Vgs.

## 12.5 BIPOLAR JUNCTION TRANSISTOR (BJT)

The bipolar junction transistor, or BJT, model in PSpice is an enhanced version of the Gummel-Poon model. This means that it is also a superset of the earlier Ebers-Moll model, as well as its more basic form which is usually the first model encountered by an electrical student. You have access to all of these levels of model by the way the Gummel-Poon parameters are set, or defaulted. Associated with this DC model are all of the junction capacitances, which, with some care, give good AC and transient simulation results up to microwave fequencies.

Both the Ebers-Moll and Gummel-Poon models are symmetrical, with both forward and reverse operation (just like a ''real'' bipolar transistor). Therefore, there are forward and reverse parameters that are explicitly labeled as such; however, some of the parameters labeled as being associated with the base-emitter or base-collector junction are also forward or reverse parameters (respectively). This means that of the forty-odd parameters in the bipolar model, most of them are duplicates specifying reverse operation, or base-collector instead of base-emitter characteristics. So, the list of parameters is not as formidable as it looks.

Using graduated models is a common way to teach transistor theory; this is easy to do since these models trace the development of the theory of transistor operation. First came the simple, nonlinear model described by Ebers and Moll, in 1954, which was a DC model only (that is, it did not include capacitive effects). This is the model you get in SPICE using the default values for the BJT parameters; the forward beta (BF) is 100, the reverse beta (BR) is 1, and IS is set to provide a normal base-emitter voltage for a small device.

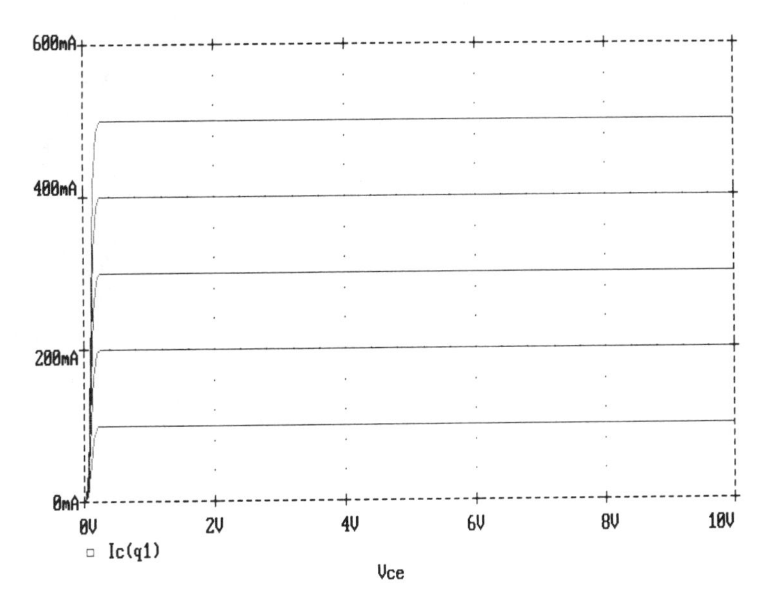


**Figure 12.13** Plot of collector current ''curve family'' for the bipolar transistor.

To get to the next level of model, you would include the junction capacitances and parasitic resistances for each of the terminals. The capacitance models are identical to the ones we saw earlier for the diode, applied to both the base-emitter and base-collector junctions; these provide correct transient/frequency response and include diffusion charge to model switching times correctly. This includes the depletion capacitance parameters:

CJE and CJC, which are equivalent to CJO for the diode
VJE and VJC, which are equivalent to VJ for the diode
MJE and MJC, which are equivalent to M for the diode
FC, which is for b-e and b-c junctions, and equivalent to FC for the diode

This also includes the diffusion capacitance parameters, TF and TR, which are equivalent to TT for the diode.

The parasitic resistances model the bulk resistance included in the physical construction of the device. The emitter (RE) and collector (RC) resistances alter the terminal characteristics, decreasing the slope of collector current for low collector-emitter voltages. The base (RB) resistance primarily affects frequency response and noise.

The final level of model includes carrier recombination and base-width modulation effects which provide the realities of gain variation. These effects are associated with the Gummel-Poon model, although included in enhanced versions of Ebers-Moll models, because the Gummel-Poon model treated a number of effects in a unified manner. The terminal characteristics were not that much different between the Gummel-Poon and earlier "enhanced" Ebers-Moll models, but the underlying physics was "better." The SPICE user must remember that most of simulation modeling is a curve fitting game; a variety of approaches will give the same (for engineering purposes) simulation results.

Base-width modulation comes from the voltage across the base-emitter and base-collector junctions. The most obvious effect is a finite output conductance, or an increase in collector current with base-collector voltage, which was called the "Early" effect (after J. M. Early, who first reported the phenomenon). The forward parameter is the Early voltage, VAF (or VA), from the following geometric interpretation: extrapolating the collector currents, in saturation, forms a converging set of lines that intersect the negative X-axis at the Early voltage (which, however, is expressed as a positive value). The output conductance, 1/hoe, is the slope of the extrapolated lines.

There is also a reverse-Early voltage, VAR (or VB), sometimes called the "Late" voltage. This parameter has the same effect, and geometric interpretation, for reverse transistor operation (see Fig. 12.14).

Carrier recombination and leakage accounts for the decrease in current gain at low current levels. Not all of the current flowing through the base terminal is available for transistor action; some of it leaks off and some is lost to recombination.

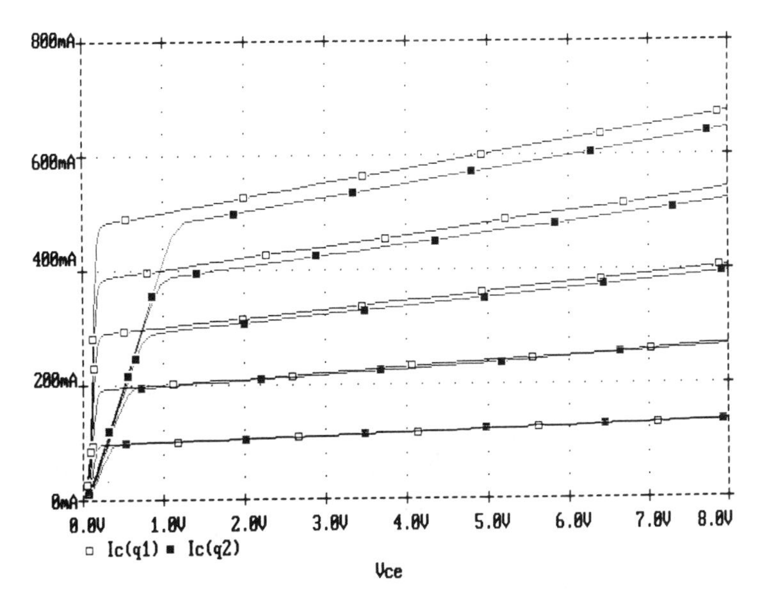


**Figure 12.14** Plot of collector current "curve families," including parasitic resistance and Early voltage.

Only the current that escapes these effects participates in the amplification action of the transistor. Leakage and recombination currents have voltage dependencies similar to the Shockley equation:

$$\text{leakage current} = IS_L \cdot (e^{V/(4 \cdot Vt)} - 1)$$

$$\text{recombination current} = IS_R \cdot (e^{V/(2 \cdot Vt)} - 1)$$

These currents are part of the base junction current, and similar currents occur in other semiconductor junctions. For example, the semiconductor diode also has these currents, but they are not modeled in SPICE because the effects do not matter for most circuits; they do matter for more complete modeling of the bipolar transistor.

As a simplification, the leakage and recombination currents are combined into a single formula:

$$\text{composite ("lost") current} = ISE \cdot (e^{V/NE \cdot Vt)} - 1)$$

where ISE and NE are the values used that make the composite formula match the combination of the previous formulas. The composite formula is the Shockley equation for a nonideal, or "leakage," diode.

In the Gummel-Poon model, this leakage diode is connected in parallel with an ideal diode to represent the base-emitter junction. The current through the ideal diodes takes part in the transistor action (its current is multiplied by beta to generate collector current); the leakage diode current does not. The parameters IS/BF and NF are the saturation current and emission coefficient, respectively, for the ideal diode. The parameters ISE and NE are the saturation current and emission coefficient, respectively, for the leakage diode. The formula for junction current, for each diode, is

$$\text{ideal diode current} = (\text{IS/BF})\cdot(e^{\text{V/(NF}\cdot\text{Vt)}} - 1)$$
$$\text{leakage diode current} = \text{ISE}\cdot(e^{\text{V/(NE}\cdot\text{Vt)}} - 1)$$

As mentioned earlier, these currents also occur in the semiconductor diode. If you wanted to model a diode more accurately in the low-current, forward region then you might try using this dual-diode approach.

Forward current gain is defined as the ratio of collector current to base current. With the leakages in the nonideal diode having an emission coefficient, NE, around 2 (the ideal diode's emission coefficient, NF, is usually unity) the percentage of total current "leaking" increases with decreasing current (both diodes have the same voltage across their terminals). The remaining current available for transistor action decreases with decreasing current and along with it the apparent beta. Here, beta (the BF parameter) is constant but the amount of current in the ideal diode is decreasing.

Forward current gain will be reduced to half of BF when these two currents are equal, and since they are in parallel the junction voltages are identical

$$\text{I @ half-beta}' = e^{(\text{NE}\cdot\ln(\text{ISE}) - \text{NF}\cdot\ln(\text{IS/BF}))/(\text{NE}-\text{NF})}$$

and the asymptotic slope of the reduction in forward gain, with decreasing current, is

$$d\text{HFE}/d(ln(\text{Ic})) = 1 - \text{NF/NE}$$

Curve fitting is used to match device measurements to these equations.

We can explore the effects of ISE and NE by using Probe to display base and collector currents. If we sweep an injected base current (using the DC sweep) and display the logarithm of base and collector current with respect to the logarithm of the ideal diode current, then we can see how forward beta changes over a broad range of forward operation.

Sweeping the base current and the value for ISE, we get Figure 12.15. In this figure the X-axis has been set to the value of the collector current. Now we can plot Ic and Ib, then label the intercepts of the trace to indicate various model parameters. Note that both axes are logarithmic. The vertical distance between the base and collector currents is the logarithmic value of the DC beta. So we can see that ISE sets the onset of reducing beta.

To see this more clearly, in Figure 12.16 we display DC beta directly as Ic/Ib. The onset of beta reduction is very clear, but it is now more difficult to relate the curves to the model parameters IS and ISE. Similarly, we may sweep base current and value for NE (the emission coefficient). Again, the X-axis has been set to Ic but with a correction factor; specifically, the diodes which model the reverse character-

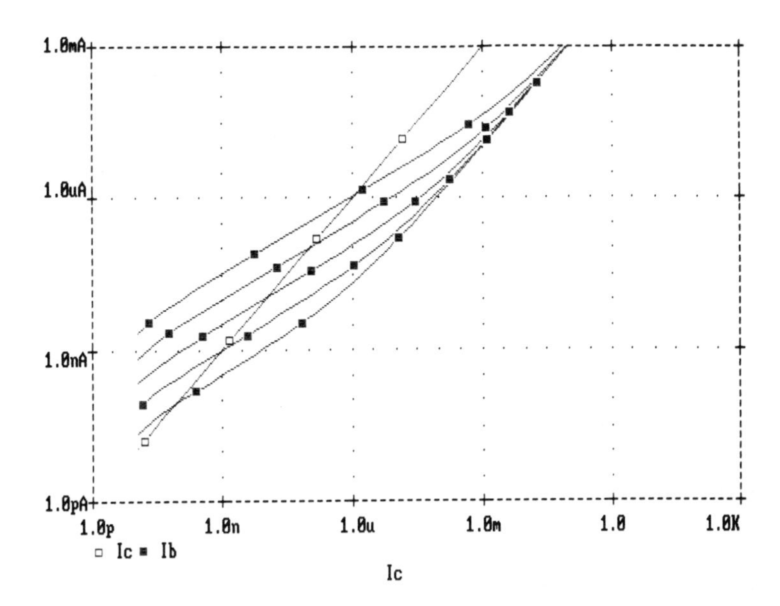


**Figure 12.15** Plot of transistor currents, varying ISE parameter.

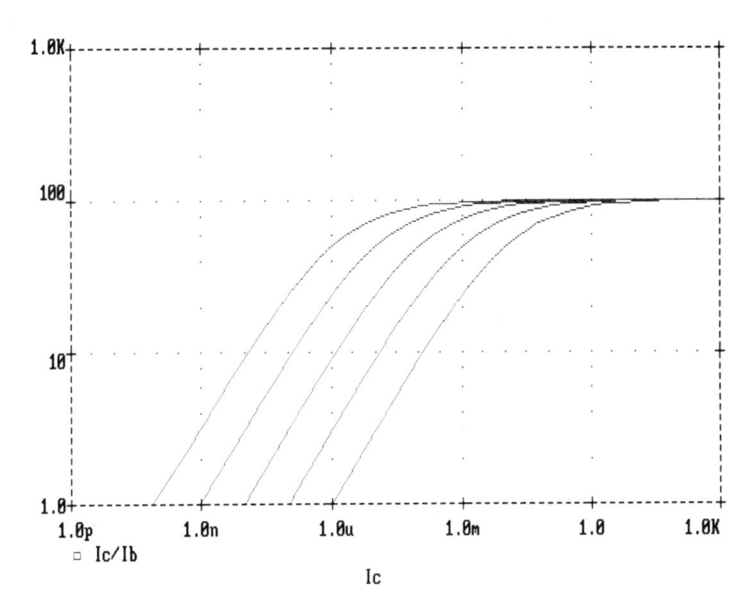


**Figure 12.16** Plot of transistor DC beta, varying ISE parameter.

istics of the transistor have currents which must be accounted for. Similarly, the traces for Ic and Ib have corrections. This is done in Figure 12.17 as the traces are shown down to extremely low currents where these corrections are significant. The traces for base current converge at the value of the model parameter ISE with an asymptotic slope of 1/NE. Again, the distance between the base and collector currents is the logarithm of the DC beta. The value of NE sets the rate at which DC beta decreases with decreasing collector current.

Of course, similar parameters are available for reverse operation: IS/BR is the reverse saturation current for the reverse "ideal" diode; ISC and NR are the saturation current and emission coefficient, respectively, for the reverse "leakage" diode.

Finally, the base-width modulation effects also account for the reduction in current gain with increasing collector current, a mode called "high-level injection." Charge conservation reduces the efficiency of the transistor action, with high-current beta having a dependence on collector current of

$$\text{beta} = \text{BF}/(1 + \text{Ic}/\text{IKF})$$

where IKF is the forward "knee" current. Solving this formula to find the collector current yielding beta equal to BF/2 shows that this occurs at a collector current equal to the value of IKF. This time the X-axis has been set to Vbe; this is proportional to the logarithm of the base (ideal diode) current and is therefore proportional to the logarithm of the "ideal" collector current (which we used in the previous figures). If we had set the X-axis to Ic, then the trace of Ic would be a straight line, as in the previous figures and we would not see the deviation of Ic due to high injection.

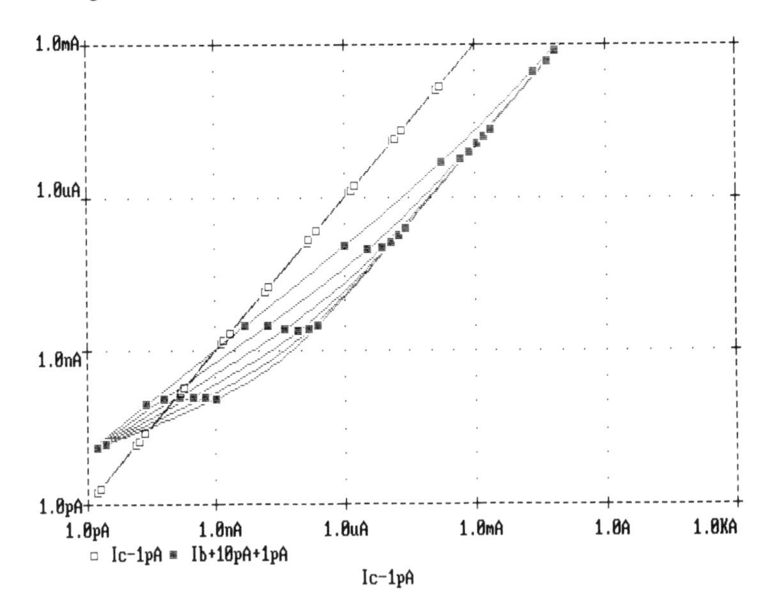


**Figure 12.17** Plot of transistor currents, varying NE parameter.

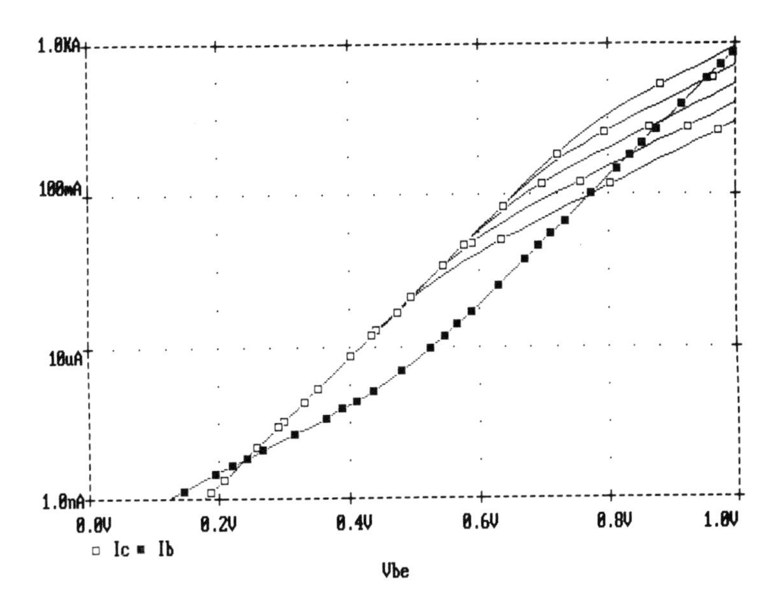


**Figure 12.18** Plot of transistor currents, varying IKF parameter.

Since DC beta is the ratio of Ic to Ib, the vertical distance between Ic and Ib in Figure 12.18 is the logarithm of DC beta. We can see that at low currents, DC beta is unity (where the traces cross) and increases to a maximum, then decreases to unity at high currents (where the traces cross again). IKR is available to model reverse operation in the same way.

This completes our internal model of the bipolar transistor for Gummel-Poon, although further improvements have been made to extend the usefulness of the model. In particular, the model includes: (i) variable base resistance, to provide for base current "crowding" effects, (ii) split base-collector capacitance, to model more accurately high-frequency response, and (iii) a variable, forward transit time to reduce frequency response at high collector currents.

## 12.6 MOS FIELD EFFECT TRANSISTOR (MOSFET)

The MOS modeling techniques of SPICE2 (and PSpice) were a significant improvement over those in the original SPICE, as well as most other simulators. The MOS models, and the availability of SPICE2 in the public domain, made the simulator popular with integrated circuit designers worldwide; these two features probably clinched the "de facto standard" title for SPICE2. But being widely used does not mean the models were widely appreciated. A cottage industry grew up surrounding SPICE2 primarily to modify the MOS models and their interaction with the circuit-solution algorithms. Nearly every integrated circuit manufacturer has an employee, or group, which supports an internal version of SPICE2. Whereas SPICE2 added two levels of MOS model to the original SPICE, some commercial programs, such as HSPICE, have an additional eighteen levels of MOS model.

This dissatisfaction with the U.S. Berkeley MOS models stems, in part, from a lack of documentation of the models themselves. There is one laboratory report, from U.C. Berkeley, which describes the models (as planned) and, of course, there is the model code (as built)—over two thousand lines of FORTRAN that do the calculations. In many cases, these sources have not been enough to use the models successfully. We will cover the formulations and parameters required to specify a model, without going into the mathematics that are covered in the U.C. Berkeley report (see the references at the end of this chapter).

There are three levels of model in SPICE2 (and available in PSpice):

Level 1: ''Schichman-Hodges'' is a basic MOSFET model and has equations that are quite similar to the JFET model.

Level 2: ''analytic model'' is a geometry-based model that attempts to calculate all effects from detailed device physics.

Level 3: ''empirical model'' is a more qualitative model that uses (as the name would imply) observed operation to define its equations.

The arguments over MOS models usually happen over levels 2 and 3; level 1 is normally used for large devices (discrete parts, such as signal MOSFETs and power MOSFETs) or for the ''first pass'' at an integrated circuit design to check that the circuit is connected and functionally correct.

The MOS models were set up for considerable flexibility in the use of parameters. While conceptually it is convenient to separate each model from the others, the equations in the code do not make this distinction for the entire model. For each characteristic, a selection of ways are available for calculating the model: level 1 is elementary, level 2 uses processing parameters and geometry, level 3 uses measured characteristics. For example, the level 2 method calculates threshold voltage from the specification of doping concentration, surface state density, et cetera, so the overall model accuracy depends on having good formulas and accurate data. However, the level 3 method uses a measured value for the threshold voltage, so the overall model accuracy depends on the ability of the engineer to match the characteristics of the component with the parameters of the model. You can use mix methods; for example, you could choose the level 2 technique for calculating threshold voltage, and the level 3 technique for calculating drain current. It would be more accurate to say that there are three methods, instead of models, as all of the combinations of the modeling sections provide many unique paths for calculating device characteristics.

Which method is used is determined, in part, by the model parameters specified. For example, if the substrate doping is specified the analytic model will be used for some calculations regardless of level. This selective calculation makes it possible to use the empirical model even though the parameter based on measured data is not available, by calculating the parameter from process data. This approach tries to arrive at a consistent set of parameters for the model equations. When enough parameters are not supplied the simulator uses default values which will, at least, provide a computable model, although probably not the model you wanted.

Level 1 is simple, like the JFET where the increase in gate-junction bias attracts charge to form the channel and modulates its resistance to drain current. It is known as a "square law" device because of the following expression relating drain current to gate-junction voltage:

$$\text{Id} = (\text{width/length}) \cdot (\text{KP}/2) \cdot (\text{Vgs-Vto})^2$$

See Fig. 12.19. Unlike the JFET, the transconductance parameter, KP, relates device size to drain current. As you may recall, with the JFET it was necessary to approximate the gate capacitance as a linear function of the gate-junction voltage (which in turn describes how the channel region is modulated). For the MOS transistor the capacitance is set, substantially, by the thickness of the gate oxide and the area of the gate, neither of which vary. This forms a linear, and nearly perfect capacitor; the same materials are used in some memory devices that store charge for a useful life measured in decades.

The square law result applies only when Vds is greater than Vgs-Vto (the "pinch-off" voltage), when the channel of the MOSFET is "saturated." When Vds is below pinch-off, the expression relating drain current to gate-junction voltage is

$$\text{Id} = (\text{width/length}) \cdot (\text{KP}/2) \cdot (2 \cdot (\text{Vgs-Vto}) \cdot \text{Vds} - \text{Vds}^2)$$

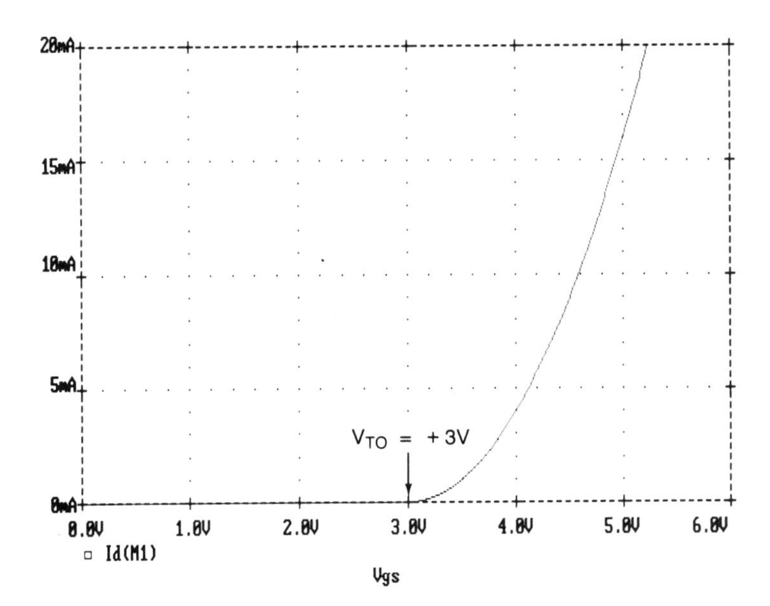


**Figure 12.19** Plot of drain current versus gate-source voltage.

which describes (on an Id versus Vds plot) an inverted parabolic curve passing through the origin and, at its peak value (when Vds is at pinch-off), intersects the square law formula. This parabolic region of operation is called the "linear" region; for small drain voltages, the expansion of the equation (above) is dominated by the linear term

$$Id \approx (width/length) \cdot KP \cdot (Vgs\text{-}Vto) \cdot Vds$$

Finally, Ids is zero when Vgs is less than Vto (see Fig. 12.20).

Nonlinear capacitance models are available for the MOSFET regardless of model level. Semiconductor *pn* junction capacitance between the substrate (bulk) and source, or drain, is modeled the same way as for the diode (except that diffusion capacitance is not included as these junctions are normally reversed-biased). Overlap capacitance, the excess overlap of the gate over any of the other sections of the device (due to the manufacturing process), is modeled as a fixed, stray capacitance to be added to any other calculated values. Overlap capacitance may be specified for any level of the MOS model. The remaining capacitance to be calculated is due to the operation of the intrinsic MOSFET, that is, these capacitance values come from the electrical characteristics of the charges in the channel and not the physical implementation of the device.

There are two models for the channel charge related capacitances: (i) the Meyer model, which empirically splits the total capacitance into varying amounts between the gate and any other terminal, and (ii) the Ward-Dutton model (available only for the level 2 model), which calculates the distribution of charge and uses a three-terminal, nonreciprocal capacitor model. For the level 2 model, the XQC

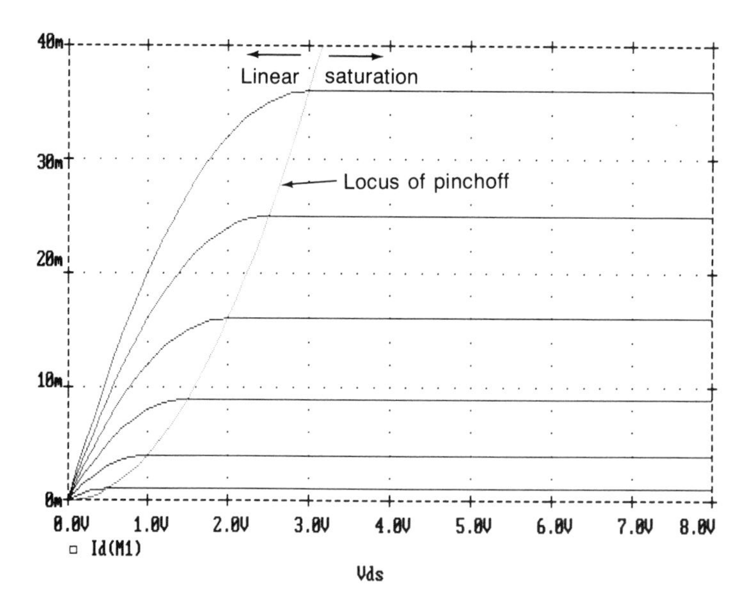


**Figure 12.20** Plot of drain current "curve family," with locus of pinch-off voltage.

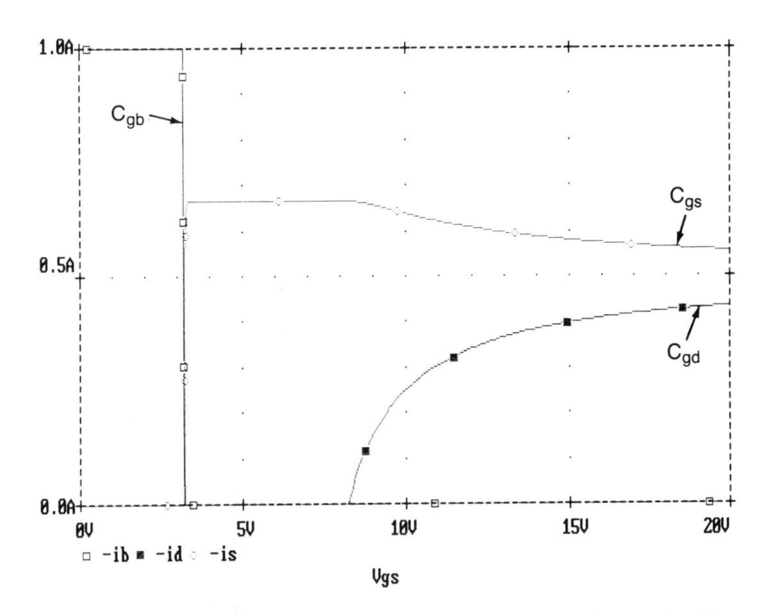


**Figure 12.21** Plot of MOS capacitance versus gate-source voltage, using Meyer formulation.

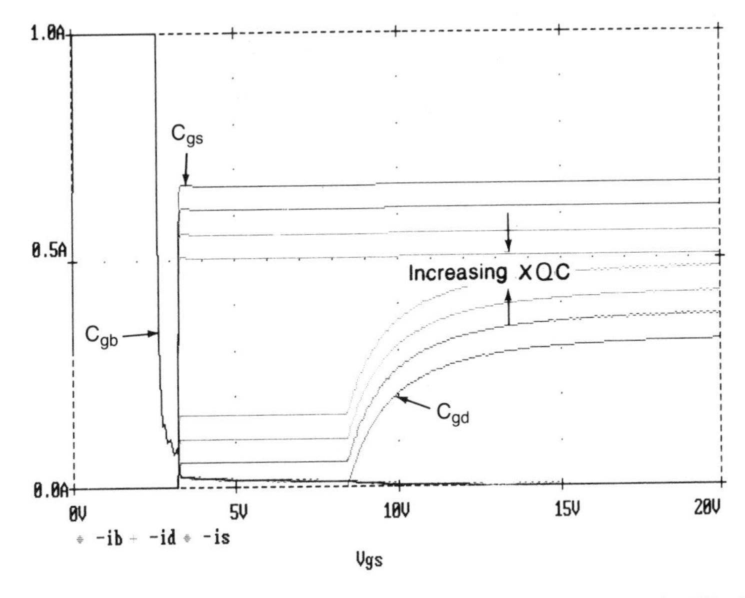


**Figure 12.22** Plot of MOS capacitance versus gate-source voltage, using Ward-Dutton formulation.

parameters selects which model is used: if XQC has a value greater than 0.5 the Meyer mode is used, otherwise the Ward-Dutton model is used. **Both models conserve charge.** The charge conservation rumors are due to a bad reputation acquired by an earlier version of SPICE2. Charge conservation is a problem of numerical integration and independent of the theory of underlying capacitance models.

## 12.7 NONLINEAR MAGNETICS

MicroSim developed a nonlinear magnetics device based on the Jiles-Atherton magnetics model. This model is based on existing ideas of domain wall motion, including flexing and translation, to simulate the behavior of the magnetic material and thereby generate B-H curves. The slope of the B-H curves then set the inductance and current values for the windings associated with the magnetic core. The model accounts for the following nonlinear effects: initial permeability, saturation of magnetization, hysteresis (including coercivity and remanence), and dynamic core losses.

The Jiles-Atherton model supposes that the magnetic material is made up of loosely coupled domains which have an equilibrium B-H curve, called the "anhysteric." This curve is the locus of B-H values generated by superimposing a DC magnetic bias and a large AC signal which decays to zero. It is the curve representing minimum energy for the domains and is modeled, in theory, by

$$\mathrm{Ma} = \mathrm{Msat} \cdot F(\mathrm{Heff}/\mathrm{a})$$

where $\mathrm{Heff} = \mathrm{H} + \mathrm{alpha} \cdot \mathrm{M}$

$$F(x) = coth(x) - 1/x$$

and Ma is the anhysteric magnetization

Msat is the saturation magnetization

Heff is the effective magnetizing influence

a is a shape parameter

alpha is a field parameter

For a given H (magnetizing influence) the anhysteric magnetization is the global flux level the material would attain if the domain walls could move freely. Instead the walls are stopped, or pinned, on dislocations in the material. The wall remains pinned until enough magnetic potential is available to break free and travel to the next pinning site. The theory supposes a mean energy required, per volume, to move domain walls. This is analogous to mechanical "drag." A (simplified) equation of this is

$$\textit{change in magnetization} = \text{potential/drag}$$

or

$$dM/dH = (Man - M)/k$$

where k is the pinning energy per volume (drag)

So much for irreversible domain wall motion. Reversible wall motion comes from flexing in the domain wall, especially when it is pinned at a dislocation, due to the magnetic potential (that is, the magnetization is not the anhysteric value). The theory supposes spherical flexure to calculate energy values and arrives at the (simplified) equation

$$dM/dH = C \cdot d(Man - M)/dH$$

which must be added to the previous state equation. In Figure 12.23 you will see both the major B-H loop (where the magnetization is brought near to the positive and negative saturation value of the material) and some minor loops (where the magnetization is varied about an offset value). The locus of B versus H depends on the history of the material and does not follow the same path the way, say, a diode's DC current follows its DC bias voltage. This is one of the reasons that make magnetic materials difficult to model: there is not a single, explicit equation for B versus H.

Magnetic materials are used in inductor and transformer cores to provide high values of inductance in a small volume and to ''trap'' the majority of the magnetic flux within the windings for efficient energy transfer. Unfortunately, the materials are also nonlinear, which means that since the value of $dB/dH$ (the slope of the curve) is proportional to the inductance of the component using the material as its

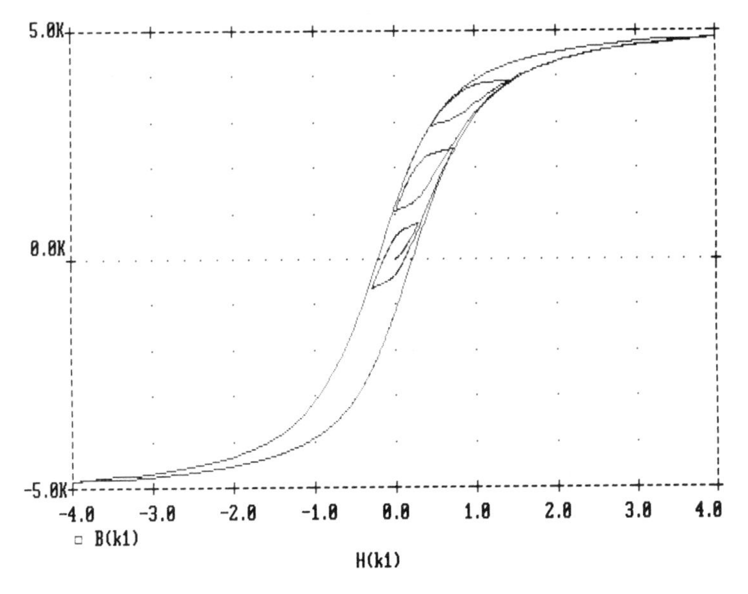


**Figure 12.23** Plot of B-H curve showing major and minor loops (ferrite material).

core, the inductance therefore varies with the current through the windings. This effect can be seen more vividly by directly displaying, with Probe, the slope of the major loop. Notice how the trace starts at a low value, at H = 0; this value is proportional to the initial inductance of the material (see Fig. 12.24).

Air gaps are available in the model. If the gap thickness is small compared with the other dimensions of the core, we can assume that all of the magnetic flux lines will go through the gap directly and that there will be little "fringing flux" (having a modest amount of fringing flux will only increase the effective air-gap length). In checking the field values around the entire magnetic path, we arrive at the equation,

$$\text{Hcore} \cdot \text{Lcore} + \text{Hgap} \cdot \text{Lgap} = \text{n} \cdot \text{I}$$

where n·I is the sum of the amp-turns of the windings on the core. Also, we know that the magnetization in the air gap is negligible so that Bgap = Hgap, and that Bgap = Bcore. These combine in the previous equation to yield

$$\text{Hcore} \cdot \text{Lcore} + \text{Bcore} \cdot \text{Lgap} = \text{n} \cdot \text{I}$$

This is a difficult equation to solve, especially for the Jiles-Atherton model, which is a state equation model rather than an explicit function (which you would expect since the B-H curve depends on the history of the material). However, there is a

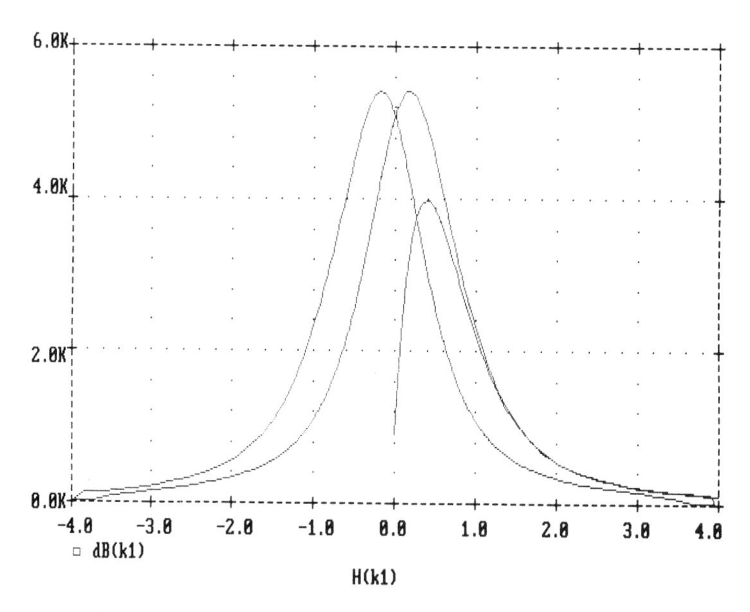


**Figure 12.24** Plot of dB/dH, or permeability, for major loop.

graphical technique that solves for Bcore and Hcore, given n · I, which is to: (i) take the nongapped B-H curve, (ii) extend a line from the current value of n · I (on the H-axis) with a slope of -Lcore/Lgap (this would be vertical if Lgap = 0), then (iii) find the intersection of the line with the B-H curve. The intersection point is the value for Bcore and Hcore for the n · I of the gapped core. The n · I value is the apparent, or external, value of Hcore, but the real value of Hcore is less. This results in a smaller value for Bcore and the "sheared over" B-H curves of a gapped core. PSpice implements the numerical equivalent of this graph technique.

The resulting B-H values are recorded in the Probe data file as Bcore and Happarent, since this is what the circuit "sees."

## 12.8 REFERENCES

For SPICE users who want to know more about the models they are using and how they relate to the component operation, I suggest the following texts:

P. Antognetti, and G. Massobrio, *Semiconductor Device Modeling with SPICE*, McGraw-Hill Company.

This new reference is a very comprehensive guide to the physics, and the derivation of the equations, for the devices in SPICE2. If you can buy only one of the items in this list, get this one.

David A. Hodges and Horace G. Jackson. *Analysis and Design of Digital Integrated Circuits*. McGraw-Hill Company.

This book has excellent sections on diode, bipolar transistor, MOSFET, and (in the latest edition) GaAsFET models and how they relate to digital IC design. Examples and exercises are given for both hand calculations and SPICE simulation.

Ian Getreu, *Modeling the Bipolar Transistor,* Tektronix, Inc., part # 062–2841–00.

This is the standard reference for a concentrated look at the development of bipolar transistor models. The only shortcoming is the book was printed in 1976, before SPICE2 was released, so the model development does not include the extensions of the SPICE2 bipolar transistor.

A. S. Grove, *Physics and Technology of Semiconductor Devices,* John Wiley & Sons, Inc.

S. M. Sze, *Physics of Semiconductor Devices*, John Wiley & Sons, Inc.

The titles give these away. More physics are included in these books, which are the standard semiconductor physics references. These texts focus on device operation and only mention circuitry and uses in passing. While there are no references to circuit simulation, all the derivations for the formulas SPICE uses, like the Shockley equation, are included.

D. C. Jiles, and D. L. Atherton, "Theory of ferromagnetic hysteresis," *Journal of Magnetism and Magnetic Materials*, 61, 48 (1986).

For those who want the physics behind the magnetics model in PSpice.

Andrei Vladimirescu and Sally Liu, "The Simulation of MOS Integrated Circuits Using SPICE2," Memorandum #M80/7.

This document describes, in detail, the MOS levels 2 and 3 device equations. Get this only if you are serious, as the text is fairly terse and knowledge of MOS device physics is assumed. This document is available by sending a check for $10 payable to *The Regents of the University of California* to this address:

Cindy Manly
EECS/ERL Industrial Liaison Program
497 Cory Hall
University of California
Berkeley, California 94720

B. J. SHEU, D. L. SCHARFETER, P. KO, and M. JENG, ''BSIM: Berkeley Short-Channel IGFET Model for MOS Transistors,'' *IEEE Journal of Solid-State Circuits,* vol. 22, no. 4, pp. 558–66 (1987).
This paper describes the new ''level 4'' MOS model in SPICE3. This model is reported to work well for small devices and has a parameter set which may be automatically extracted using semiconductor measurement equipment.

# APPENDIX A

# Abridged Summary of PSpice Statements

This section will quickly review control statements available in PSpice as of late 1987. Each statement is simply described by its use in the circuit file, with some comments on its use. More detailed comments are in the pages of this text, as well as the PSpice User's Guide (available from MicroSim Corporation).

## *  **Comment**

General forms

*(any text)

*Examples*

```
*This is an example of a comment
```

A statement beginning with "*" is a comment line and has no effect. The use of comment statements throughout the input is recommended. For example, it is good practice to place a comment just before a subcircuit definition to identify the nodes:

```
*              +IN -IN V+ V-  +OUT -OUT
.SUBCKT OPAMP 100 101 1  2   200  201
```

### ; In-line Comment

General forms

(circuit file text); (any text)

*Examples*

```
R13  6 8 10K ;feedback resistor
C3  15 0 .1U ;decouple supply
```

A ";" is treated as the end of a line so that PSpice moves on to the next line in the circuit file. The text after the ";" is a comment and has no effect.

### .AC AC Analysis

General forms

.AC [LIN][OCT][DEC] <*points value*>
+ <*start frequency value*> <*end frequency value*>

*Examples*

```
.AC LIN 101 100Hz  200Hz
.AC OCT  10    1KHz  16KHz
.AC DEC  20    1MEG 100MEG
```

The .AC statement is used to calculate the frequency response of a circuit over a range of frequencies. LIN, OCT, or DEC are keywords that specify the type of sweep, and <*points value*> the number of points in the sweep:

- LIN Linear sweep. The frequency is swept linearly from the starting to the ending frequency. <*points value*> is the total number of points in the sweep.
- OCT Sweep by octaves. The frequency is swept logarithmically by octaves. <*points value*> is the number of points per octave.
- DEC Sweep by decades. The frequency is swept logarithmically by decades. <*points value*> is the number of points per decade.

### .DC DC Analysis

General forms

.DC [LIN] <*sweep variable name*>
+ <*start value*> <*end value*> <*increment value*>

+ [*nested sweep specification*]
.DC [OCT] [DEC] <*sweep variable name*>
+ <*start value*> <*end value*> <*points value*>
+ [*nested sweep specification*]
.DC <*sweep variable name*> LIST <*value*>*
+ [*nested sweep specification*]

*Examples*

```
.DC VIN -.25 .25 .05
.DC LIN I2 5mA -2mA 0.1mA
.DC VCE 0V 10V .5V IB 0mA 1mA 50uA
.DC RES RMOD(R) 0.9 1.1 .001
.DC DEC NPN QFAST(IS) 1E-18 1E-14 5
.DC TEMP LIST 0 20 27 50 80 100 -50
```

The .DC statement causes a DC sweep analysis to be performed on the circuit. The DC sweep analysis calculates the circuit's bias-point over a range of values for <*sweep variable name*>. See Chapter 4 for use of this type of analysis.

A nested sweep is available. A second sweep variable, sweep type, start, end, and increment values may be placed after the first sweep. In this case the first sweep will be the "inner" loop: the entire first sweep will be done for each value of the second sweep. The rules for the values in the second sweep are the same as for the first.

The sweep can be linear, logarithmic, or a list of values. If linear, the keyword LIN is optional. The sweep type can be:

LIN Linear sweep. The sweep variable is swept linearly from the starting to the ending value. <*increment value*> is the step size.

OCT Sweep by octaves. The sweep variable is swept logarithmically by octaves. <*points value*> is the number of steps per octave.

DEC Sweep by decades. The sweep variable is swept logarithmically by decades. <*points value*> is the number of steps per decade.

LIST Use a list of values. In this case there are no start and end values. Instead, the numbers that follow the keyword LIST are the values that the sweep variable will be set to.

<*sweep variable name*> can be one of the following types:

**Source:** a name of an independent voltage, or current, source. During the sweep the source's voltage or current is set to the sweep value.

**Model parameter:** a model type and model name followed by a model parameter name in parenthesis. The parameter in the model is set to the sweep value.

**Temperature:** use the keyword TEMP for *<sweep variable name>*. The temperature is set to the sweep value. For each value in the sweep all the circuit components have their model parameters updated to that temperature.

## .END End of Circuit

General forms

.END

*Examples*

```
.END
```

The .END statement marks the end of the circuit. All the data and commands must come before it. When the .END statement is reached, PSpice does all the specified analyses on the circuit.

There may be more than one circuit in an input file. Each circuit and its commands are marked by a .END statement. PSpice processes all the analyses for each circuit before going on to the next one. Everything is reset at the beginning of each circuit. Having several circuits in one file gives the same results as having them in separate files and running each one separately. This is a convenient way to arrange a set of runs to be done overnight.

## .ENDS End of Subcircuit Definition

General forms

.ENDS [*subcircuit name*]

*Examples*

```
.ENDS
.ENDS  OPAMP
```

The .ENDS statement marks the end of a subcircuit definition (started by a .SUBCKT statement). It is good practice to repeat the subcircuit name although this is not required.

## .FOUR Fourier Analysis

General forms

.FOUR *<frequency value>* *<output variable>**

*Examples*

```
.FOUR 10KHz V(5) V(6,7) I(VSENS3)
```

Fourier analysis performs a decomposition into Fourier components of the result(s) of a transient analysis. See Chapter 10 for use of this type of analysis.

## .IC Initial Transient Conditions

General forms

.IC < V(<*node*>) = <*value*> >*

*Examples*

```
.IC V(2)=3.4 V(102)=0 V(3)=1V
```

The .IC statement is used to set initial conditions for transient analysis. Each <*value*> is a voltage which is assigned to <*node*> for the duration of the bias-point calculation for the transient analysis. After the bias-point has been calculated and the transient analysis started, the node is "released."

The .IC sets initial conditions for the transient analysis only. It does not affect the regular bias-point calculation or the DC sweep.

## .INC Include File

General forms

.INC [*file name*]

*Examples*

```
.INC SETUP.CIR
.INC C:\LIB\VCO.CIR
```

The .INC statement is used to insert the contents of another file. Included files may contain any statements with the following exceptions: no title line is allowed (use a comment), .END statement (if present) marks only the end of included file, .INC statement may be used (only up to four levels of "including").

Including a file is the same as simply bringing the file's text into the circuit file. Everything in the included file is actually read in, and every model and subcircuit definition, even if not needed, takes up space in main memory (RAM).

## .LIB Library File

General forms

.LIB [*file name*]

*Examples*

```
.LIB
.LIB OPNOM.LIB
.LIB C:\LIB\QNOM.LIB
```

The .LIB statement is used to reference a model or subcircuit library in another file. If [*file name*] is left off it defaults to "NOM.LIB." Library files may contain comments, .MODEL statements, subcircuit definitions (including the .ENDS statement), and .LIB statements. No other statements are allowed.

Referencing a library is not the same as simply bringing the file's text into the circuit file. Only those model or subcircuit definitions which are called by the circuit file are actually read in. So, only those model or subcircuit definitions which are needed take up space in main memory (RAM).

## .MC Monte Carlo Analysis

General forms

.MC <*#runs value*> [DC][AC][TRAN] <*output variable*> YMAX
+ [LIST][OUTPUT <*output specification*>

*Examples*

```
.MC 10 TRAN V(5) YMAX
.MC 50 DC IC(Q7) YMAX LIST
.MC 20 AC VP(13,5) YMAX LIST OUTPUT ALL
```

The .MC statement causes a Monte Carlo (statistical) analysis of the circuit. Multiple runs of the selected analysis (DC, AC, transient) are done. The first run is done with nominal values of all components. Subsequent runs are done with variations on model parameters as specified by the DEV and LOT tolerances on each .MODEL parameter (see the .MODEL statement for details on the DEV and LOT tolerances). <*#runs value*> is the total number of runs to do. The other specifications on the .MC statement control the output generated by the Monte Carlo analysis.

Exactly one of DC, AC, or TRAN must be specified. This analysis will be repeated in subsequent passes of the analysis. All analyses that the circuit

contains are performed during the nominal pass. Only the selected analysis is performed during subsequent passes.
*<output variable>* is identical in format to that of a .PRINT output variable. The keyword YMAX specifies the operation to be performed on the values of the *<output variable>* to reduce these to a single value. This value is the basis for the comparisons between the nominal and subsequent runs. YMAX is the only reduction method currently implemented. Others will be added as user requirements indicate.

If the keyword LIST is specified PSpice will print out, at the beginning of each run, the model parameter values actually used for each component during that run.

The output from the nominal (first) run is governed by the .PRINT, .PLOT, and .PROBE statements in the file. The output of subsequent runs are suppressed unless requested by the OUTPUT keyword: *<output specification>* is one of the following:

| | |
|---|---|
| ALL | forces all output to be generated |
| FIRST *<value>* | generates output only during first n runs |
| EVERY *<value>* | generates output every nth run |
| RUNS *<value>** | generates output only for the listed runs |

## .MODEL Model

General forms

.MODEL *<name>* *<type name>*
\+ ([*<parameter name>* = *<value>* [*tolerance specification*]]*)

*Examples*

```
.MODEL RMAX RES (R=1.5 TC1=.02 TC2=.005)
.MODEL DNOM D (IS=1E-9)
.MODEL QDRIV NPN (IS=1E-7 BF=30)
.MODEL MLOAD NMOS(LEVEL=1 VTO=.7 CJ=.02pF)
.MODEL CMOD CAP (C=1 DEV 5%)
.MODEL DLOAD D (IS=1E-9 DEV .5% LOT 10%)
```

The .MODEL statement defines a set of device parameters which can be referenced by devices in the circuit. *<name>* is the model name which devices use to reference a particular model. *<name>* must start with a letter. It is good practice to make this the same letter as the device name (e.g., D for diode, Q for bipolar transistor), but this is not required.

*<type name>* is the device type and must be one of the following:

| | |
|---|---|
| CAP | capacitor |
| IND | inductor |

| | |
|---|---|
| RES | resistor |
| D | diode |
| NPN | NPN bipolar transistor |
| PNP | PNP bipolar transistor |
| NJF | N-channel junction FET |
| PJF | P-channel junction FET |
| NMOS | N-channel MOSFET |
| PMOS | P-channel MOSFET |
| GASFET | N-channel GaAs MESFET |
| CORE | nonlinear, magnetic core (transformer) |
| VSWITCH | voltage-controlled switch |
| ISWITCH | current-controlled switch |

Devices can reference models only of the correct type. A JFET can reference a model of types NJF, or PJF, but not of type NPN. There can be more than one model of the same type in a circuit, although they must have different names.

Following <*type name*> is a list of parameter values enclosed by parenthesis. None, any, or all parameters may be assigned values. Default values are used for all unassigned parameters. The lists of parameter names, meanings, and default values are located in the individual device descriptions.

(*tolerance specification*) may be appended to each parameter, with the format

[DEV <*value*>[%]] [LOT <*value*>[%]]

These are used by the .MC analysis. LOT tolerances track, so that all devices that refer to the same model will use the same value of the model parameter. DEV tolerances are independent. The "%" indicates a relative (percentage) tolerance. If it is omitted <*value*> is in the same units as the parameter itself.

## .NODESET Nodeset

General forms

.NODESET < V(<*node*>) = <*value*> >*

*Examples*

```
.NODESET V(2)=3.4 V(102)=0 V(3)=-1V
```

The .NODESET statement helps calculate the bias-point by providing an initial guess for some nodes. Some or all of the circuit's nodes may be given an initial guess. It is effective for the regular bias-point and the bias-point for transient analysis. It has no effect during the DC sweep or during the transient analysis itself.

Unlike the .IC statement, .NODESET provides only an initial guess for some node voltages. It does not clamp those nodes to the specified voltages.

However, by providing an initial guess, .NODESET may be used to "break the tie" in, for instance, a flip-flop, and make it come up in a desired state.

## .NOISE Noise Analysis

General forms

.NOISE V(<*node*> [,<*node*>]) <*name*> [*internal value*]

*Examples*

```
.NOISE V(5) VIN
.NOISE V(101) VSRC 20
.NOISE V(4,5) ISRC
```

The .NOISE statement causes a noise analysis of the circuit to be done. Noise analysis is done in conjunction with AC analysis and requires there to be a .AC statement.

V(<*node*> [,<*node*>]) is an output voltage. It has a form such as V(5), which is the voltage at an output node, or a form such as V(4,5), which is the output voltage across two nodes. <*name*> is the name of an independent voltage or current source at which the equivalent input noise will be calculated. <*name*> is not itself a noise generator, but only a place at which to calculate the equivalent input noise.

The noise-generating devices in a circuit are the resistors and the semiconductor devices. For each frequency of the AC analysis, each noise generator's contribution is calculated and propagated to the output nodes. There, all the propagated noise values are RMS-summed. The gain from the input source to the output voltage is also computed and from it and the total output noise an equivalent input noise is calculated. If:

<*name*> is a voltage source, then the input noise units are volt/hertz$^{1/2}$

<*name*> is a current source, then the input noise units are amp/hertz$^{1/2}$

The output noise units are always volt/hertz$^{1/2}$.

If [*interval value*] is present, then it is the print interval. Every nth frequency, where n is the print interval, a detailed table is printed showing the individual contributions of all the circuit's noise generators to the total noise. These values are the noise amounts propagated to the output nodes, not the noise amounts at each generator. If [*interval value*] is not present, then no detailed table is printed.

The detailed table is printed while the analysis is being done, and does not need .PRINT or .PLOT statements. The output noise and equivalent input noise may be output with .PRINT statement or .PLOT statements if desired. Noise analysis is the only analysis for which you have a choice about using the .PRINT or .PLOT statements.

## .OP Bias-Point Analysis

General forms

.OP

*Examples*

```
.OP
```

The .OP statement causes detailed information about the bias-point to be printed. The bias-point is calculated whether or not there is a .OP statement. Without a .OP statement the only information about the bias-point which is output is a list of the node voltages.

With a .OP statement the currents and power dissipation of all the voltage sources are printed. Also the small signal (linearized) parameters of all the nonlinear controlled sources and all the semiconductor devices are output.

The .OP statement controls output for the regular bias-point only. The .TRAN statement controls output for the transient analysis bias-point.

## .OPTIONS Options

General forms

.OPTIONS [*option name*]* [ <*option name*> = <*value*> ]*

*Examples*

```
.OPTIONS NOECHO NOMOD DEFL=12u DEFW=8u DEFAD=150p DEFAS=150p
.OPTIONS ACCT RELTOL=.01
```

The .OPTIONS statement is used to set all the options, limits, and control parameters for the various analyses including the output width (see the .WIDTH statement, which is still supported).

The options are listed in any order. There are two kinds of options: those with values and those without. The options without values are flags of various kinds and simply listing the option name is sufficient.

The following table lists the flag options. The default for any flag option is "off" (i.e., the opposite of specifying the option).

| Option | Meaning |
|---|---|
| ACCT | summary and accounting information is output at the end of all the analyses (see Job Statistics Summary for further information on ACCT) |
| LIST | summary of circuit elements (devices) is output |
| NODE | net list (node table) is output |

| | |
|---|---|
| NOECHO | suppresses listing of the input file |
| NOMOD | suppresses listing of model parameters and temperature updated values |
| NOPAGE | suppresses paging and printing of a banner for each major section of output |
| OPTS | values for all options are output |
| WIDTH | Same as "WIDTH OUT=" statement |

The table below lists the options with values and their default values:

| Option | Meaning | Units | Default |
|---|---|---|---|
| ABSTOL | best accuracy of currents | amp | 1pA |
| CHGTOL | Best accuracy of charges | coulomb | .01pC |
| CPTIME | CPU time allowed for this run | sec | 1E6 |
| DEFAD | MOSFET default drain area (AD) | $meter^2$ | 0 |
| DEFAS | MOSFET default source area (AS) | $meter^2$ | 0 |
| DEFL | MOSFET default length (L) | meter | 100u |
| DEFW | MOSFET default width (W) | meter | 100u |
| GMIN | minimum conductance used for any branch | $ohm^{-1}$ | 1E–12 |
| ITL1 | DC and bias-point "blind" iteration limit | | 40 |
| ITL2 | DC and bias-point "educated guess" iteration limit | | 20 |
| ITL4 | iteration limit at any point in transient analysis | | 10 |
| ITL5 | total iteration limit for all points in transient analysis (ITL5 = 0 means ITL5 = infinity) | | 5000 |
| LIMPTS | maximum points allowed for any print table or plot | | 201 |
| NUMDGT | number of digits output in print tables (maximum eight useful digits) | | 4 |
| PIVREL | relative magnitude required for pivot in matrix solution | | 1E–3 |
| PIVTOL | absolute magnitude required for pivot in matrix solution | | 1E–13 |
| RELTOL | relative accuracy of V's and I's | | .001 |
| TNOM | default temperature (also the temperature at which model parameters are assumed to have been measured) | °C | 27 |
| TRTOL | transient analysis accuracy adjustment | | 7.0 |
| VNTOL | best accuracy of voltages | volt | 1uV |

## .PLOT Plot

General forms

.PLOT [DC] [AC] [NOISE] [TRAN] [*output variable*]*
\+ ([<*lower limit value*> , <*upper limit value*>])*

*Examples*

```
.PLOT DC V(3) V(2,3) V(R1) I(VIN) I(R2) IB(Q13) VBE(Q13)
.PLOT AC VM(2) VM(3,4) VG(5) VDB(5) IR(6) II(7)
```

```
.PLOT NOISE INOISE ONOISE DB(INOISE) DB(ONOISE)
.PLOT TRAN V(3) V(2,3) (0,5V) ID(M2) I(VCC) (-50mA,50mA)
```

The .PLOT statement allows results from DC, AC, noise, and transient analyses to be output in the form of "line printer" plots. These plots are made by using characters to draw the plot, hence they will work with any kind of printer.

DC, AC, NOISE, and TRAN are the analysis types which can be output with .PLOT statements. Exactly one analysis type must be specified.

See Chapters 4, 6, 8, and 9 for use of this statement.

## .PRINT Print

General forms

.PRINT [DC] [AC] [NOISE] [TRAN] [*output variable*]*

*Examples*

```
.PRINT DC V(3) V(2,3) V(R1) I(VIN) I(R2) IB(Q13) VBE(Q13)
.PRINT AC VM(2) VP(2) VM(3,4) VG(5) VDB(5) IR(6) II(7)
.PRINT NOISE INOISE ONOISE DB(INOISE) DB(ONOISE)
.PRINT TRAN V(3) V(2,3) ID(M2) I(VCC)
```

The .PRINT statement allows results from DC, AC, noise, and transient analyses to be output in the form of tables, referred to as print tables.

DC, AC, NOISE, and TRAN are the analysis types which can be output with .PRINT statements. Exactly one analysis type must be specified.

See Chapters 4, 6, 8, and 9 for use of this statement.

## .PROBE Probe

General forms

.PROBE

.PROBE [*output variable*]*

*Examples*

```
.PROBE
.PROBE V(3) V(2,3) V(R1) VM(2) VP(2) I(VIN) I(R2) IB(Q13)
+      VBE(Q13) VDB(5)
```

The .PROBE statement writes the results from DC, AC, and transient analyses to a data file named PROBE.DAT for use by the Probe graphics post-processor. See Chapters 4, 6, 8, and 9 for use of this statement.

The first form (with no output variables) writes all the node voltages and all the device currents to the data file. The second form writes only those output variables specified to the data file. Note that unlike the .PRINT and .PLOT statements there is no analysis name before the output variables.

## .SENS Sensitivity Analysis

General forms

.SENS *<output variable>**

*Examples*

```
.SENS V(9) V(4,3) V(17) I(VCC)
```

The .SENS statement causes a DC sensitivity analysis to be performed. By linearizing the circuit about the bias-point, the sensitivities of each of the output variables to all the device values and model parameters will be calculated and output. This can easily generate huge amounts of output.

*<output variable>* has the same format and meaning as in the .PRINT statement for DC and transient analyses. However, in the case of *<output variable>* being a current, it is restricted to be the current through a voltage source.

## .SUBCKT Subcircuit definition

General forms

.SUBCKT *<name>* [*node*]*

*Examples*

```
.SUBCKT OPAMP 1 2 101 102
```

The .SUBCKT statement begins the definition of a subcircuit. The definition is ended with a .ENDS statement. All the statements between .SUBCKT and .ENDS are included in the definition. Whenever the subcircuit is called, by an X statement, all the statements in the definition replace the calling statement.

*<name>* is the subcircuit's name and is used by an X statement to reference the subcircuit. It must start with a letter.

[*node*]* is an optional list of nodes. There must be the same number of nodes in the subcircuit calling statements as in its definition. When the subcircuit is called, the actual nodes (the ones in the calling statement) replace the argument nodes (the ones in the defining statement).

Subcircuit calls may be nested. That is, an X statement may appear between a .SUBCKT and a .ENDS.

Subcircuit definitions **may not be nested**—a .SUBCKT statement may not appear in the statements between a .SUBCKT and a .ENDS.

Subcircuit definitions should contain only device statements (statements without a leading ".") and possibly .MODEL statements. Models defined within a subcircuit definition are **available only within the subcircuit definition** in which they appear. Also, if a .MODEL statement appears in the main circuit, that model is available in the main circuit and all subcircuits.

Node, device, and model names are local to the subcircuit in which they are defined. That is, it is OK to use a name in a subcircuit which has already been used in the main circuit. When the subcircuit is expanded all its names are prefixed with the subcircuit instance name: for example, "Q13" becomes "X3.Q13." After expansion all names are unique.

## .TEMP Temperature

General forms

.TEMP *<temperature value>**

*Examples*

```
.TEMP 125
.TEMP 0 27 125
```

The .TEMP statement sets the temperature at which all analyses are done. The temperatures are in degrees Centigrade. If more than one temperature is given, then all analyses are done for each temperature. It is assumed that the model parameters were measured or derived at the nominal temperature. The nominal temperature is 27°C unless set otherwise by the TNOM option in the .OPTIONS statement.

## .TF Transfer Function

General forms

.TF *<output variable>* *<input source name>*

*Examples*

```
.TF V(5) VIN
.TF I(VDRIV) ICNTRL
```

The .TF statement causes the small-signal transfer function to be calculated by linearizing the circuit around the bias-point. The gain from *<input source name>* to *<output variable>* will be output along with the input and output resistances.

The output is done as soon these quantities are calculated and does not require .PRINT, .PLOT, or .PROBE statements.

*<output variable>* has the same format and meaning as in the .PRINT statement. However, in the case of *<output variable>* being a current, it is restricted to be the current through a voltage source.

## .TRAN Transient Analysis

General forms

.TRAN[/OP] *<print step value>* *<final time value>*
\+ [*<no-print value>* [*step ceiling value*]] [UIC]

*Examples*

```
.TRAN    1nS 100nS
.TRAN/OP 1nS 100nS 20nS UIC
.TRAN    1nS 100nS 0ns .1nS
```

The .TRAN statement causes a transient analysis to be performed on the circuit. The transient analysis calculates the circuit's behavior over time, starting at TIME = 0 and going to *<final time value>*.

The transient analysis uses an internal time step which is adjusted as the analysis proceeds. Over intervals where there is little activity, the internal time step is increased and during busy intervals it is decreased. *<print step value>* is the time interval used for printing or plotting the results of the transient analysis. Since the results are computed at different times than they are printed, a second-order polynomial interpolation is used to obtain the printed values. The transient analysis always starts at TIME = 0. However, it is possible to suppress output of a portion of the analysis. [*no-print value*] is the amount of time from Time = 0 which is not printed, plotted, or given to Probe.

Sometimes one is concerned about the size of the internal time-step. The default ceiling on the internal time-step is *<final time value>*/50 (**it is not** *<print step value>*). [*step ceiling value*] allows a ceiling smaller or larger than the print interval to be put on the internal time-step.

Prior to doing the transient analysis, PSpice computes a bias-point for the circuit separate from the regular bias-point. This is done because the independent sources can have different values at the start of a transient analysis than their DC value. Normally only the node voltages are printed for the transient analysis bias-point. However, the "/OP" suffix (on .TRAN) will cause the same detailed printing of the bias-point that the .OP statement causes for the regular bias-point.

If the keyword UIC (Use Initial Conditions) is put at the end of the .TRAN statement, the calculation of the bias-point is skipped. This option is used with the IC = specification for capacitors and inductors.

## .WIDTH Width

General forms

.WIDTH OUT = <*value*>

*Examples*

```
.WIDTH OUT = 80
.WIDTH OUT = 132
```

The .WIDTH statement sets the width of the output. <*values*> is in columns and must be either 80 (the default) or 132.

# APPENDIX B

# Abridged Summary of PSpice Devices

This section summarizes the devices available in PSpice as of late 1987. Each device is simply described by its use in the circuit file, and a list of model parameters which may be used, in the .MODEL statement, to characterize the device. The equations for device currents, capacitances, temperature corrections, and noise currents are in the PSpice User's Guide (available from MicroSim Corporation).

## B GaAsFET

General forms

B<*name*> <*drain node*> <*gate node*> <*source node*>
\+ <*model name*> [*area value*]

*Examples*

```
BIN 100 1 0 GFAST
B13 22 14 23 GNOM 2.0
```

| Model Parameters (see .MODEL statement) | | Default value | Units |
|---|---|---|---|
| LEVEL | model type (1 = Curtice, 2 = Raytheon) | 1 | |
| VTO | threshold voltage | −2.5 | volt |
| ALPHA | tanh constant | 2 | $volt^{-1}$ |

| Model Parameters (see .MODEL statement) | | Default value | Units |
|---|---|---|---|
| B | doping tail extending parameter (level 2 only) | .3 | |
| BETA | transconductance coefficient | .1 | amp/volt$^2$ |
| LAMBDA | channel-length modulation | 0 | volt$^{-1}$ |
| RG | gate ohmic resistance | 0 | ohm |
| RD | drain ohmic resistance | 0 | ohm |
| RS | source ohmic resistance | 0 | ohm |
| IS | gate p-n saturation current | 1E–14 | amp |
| M | gate p-n grading coefficient | .5 | |
| N | gate p-n emission coefficient | 1 | |
| VBI | gate p-n potential | 1 | volt |
| CGD | gate-drain zero-bias p-n capacitance | 0 | farad |
| CGS | gate-source zero-bias p-n capacitance | 0 | farad |
| CDS | drain-source capacitance | 0 | farad |
| TAU | transit time | 0 | sec |
| FC | forward-bias depletion capacitance coefficient | .5 | |
| VTOTC | VTO temperature coefficient | 0 | volt/°C |
| BETATCE | BETA exponential temperature coefficient | 0 | %/°C |
| KF | flicker noise coefficient | 0 | |
| AF | flicker noise exponent | 1 | |

The GaAsFET is modeled as an intrinsic FET with an ohmic resistance (RD/area) in series with the drain, and with another ohmic resistance (RS/area) in series with the source, and with another ohmic resistance (RG) in series with the gate. [*area value*] is the relative device area and defaults to 1.

## C Capacitor

General forms

C<*name*> <+ *node*> <– *node*> [*model name*] <*value*>
+ [IC = <*initial value*>]

*Examples*

```
CLOAD 15 0 20pF
C2  1 2 .2E-12 IC=1.5V
CFDBCK 3 33 CMOD 10pF
```

| Model Parameter (see .MODEL statement) | | Default value | Units |
|---|---|---|---|
| C | capacitance multiplier | 1 | |
| VC1 | linear voltage coefficient | 0 | volt$^{-1}$ |

| Model Parameter (see .MODEL statement) | | Default value | Units |
|---|---|---|---|
| VC2 | quadratic voltage coefficient | 0 | volt$^{-2}$ |
| TC1 | linear temperature coefficient | 0 | °C$^{-1}$ |
| TC2 | quadratic temperature coefficient | 0 | °C$^{-2}$ |

The (+) and (−) nodes define the polarity means when the capacitor has a positive voltage across it. Positive current flows from the (+) node through the capacitor to the (−) node.

If [*model name*] is left out then <*value*> is the capacitance in farads. If [*model name*] is specified, then the capacitance is given by the formula

$$\langle value \rangle \cdot C \cdot (1 + VC1 \cdot V + VC2 \cdot V^2) \cdot (1 + TC1 \cdot (T - Tnom) + TC2 \cdot (T - Tnom)^2)$$

<*value*> is normally positive (though it can be negative, but **not** zero). "Tnom" is the nominal temperature (set with TNOM option).

<*initial value*> is the initial guess for the voltage across the capacitor during the bias-point calculation.

*Noise:* The capacitor does not have a noise model.

## D Diode

General forms

D<*name*> <*+ node*> <*− node*> <*model name*> [*area value*]

*Examples*

```
DCLAMP 14 0 DMOD
D13  15 17 SWITCH 1.5
```

| Model parameter (see .MODEL statement) | | Default value | Units |
|---|---|---|---|
| IS | saturation current | 1E−14 | amp |
| N | emission coefficient | 1 | |
| RS | parasitic resistance | 0 | ohm |
| CJO | zero-bias p-n capacitance | 0 | farad |
| VJ | p-n potential | 1 | volt |
| M | p-n grading coefficient | .5 | |
| FC | forward-bias depletion capacitance coefficient | .5 | |
| TT | transit time | 0 | sec |
| BV | reverse breakdown voltage | infinite | volt |
| IBV | reverse breakdown current | 1E−10 | amp |

| Model parameter (see .MODEL statement) | | Default value | Units |
|---|---|---|---|
| EG | bandgap voltage (barrier height) | 1.11 | eV |
| XTI | IS temperature exponent | 3 | |
| KF | flicker noise coefficient | 0 | |
| AF | flicker noise exponent | 1 | |

The diode is modeled as an ohmic resistance (RS/area) in series with an intrinsic diode. <+ *node*> is the anode and <−*node*> is the cathode. Positive current is current flowing from the anode through the diode to the cathode. [*area value*] scales IS, RS, CJO, and IBV, and defaults to 1. IBV and BV are both specified as positive values.

## E Voltage-Controlled Voltage Source

General forms

E<*name*> <+ *node*> <− *node*>
\+ <+ *controlling node*> <− *controlling node*> <*gain*>
E<*name*> <+ *node*> <− *node*> POLY(<*value*>)
\+ < <+ *controlling node*> <− *controlling node*> >*
\+ < <*polynomial coefficient value*> >*

*Examples*

```
EBUFF    1   2 10 11 1.0
EAMP     13   0 POLY(1) 26 0 500
ENONLIN  100  101  POLY(2) 3 0 4 0 0.0 13.6 0.2 0.005
```

The first form and the first two examples apply to the linear case. The second form and the last example are for the nonlinear case. POLY(<*value*>) specifies the number of dimensions of the polynomial. The number of pairs of controlling nodes must be equal to the number of dimensions.

The (+) and (−) nodes are the output nodes. Positive current flows from the (+) node through the source to the (−) node. The (+ controlling) and (− controlling) nodes are in pairs and define a set of controlling voltages. A particular node may appear more than once, and the output and controlling nodes need not be different.

For the linear case, there are two controlling nodes and these are followed by the gain. For the nonlinear case see Chapter 4 for describing the controlling polynomial.

## F Current-Controlled Current Source

General forms

F<*name*> <+ *node*> <− *node*>
\+ <*controlling V device name*> <*gain*>
F<*name*> <+ *node*> <−*node*> POLY(<*value*>)
\+ <*controlling V device name*>*
\+ < <*polynomial coefficient value*> >*

*Examples*

```
FSENSE    1    2 VSENSE 10.0
FAMP      13   0 POLY(1) VIN 500
FNONLIN   100  101 POLY(2) VCNTRL1 VCINTRL2 0.0 13.6 0.2 0.005
```

The first form and the first two examples apply to the linear case. The second form and the last example are for the nonlinear case. POLY(<*value*>) specifies the number of dimensions of the polynomial. The number of controlling voltage sources must be equal to the number of dimensions.

The (+) and (−) nodes are the output nodes. A positive current will flow from the (+) node through the source to the (−) node. The current through the controlling voltage source determines the output current. The controlling source must be an independent voltage source (V device), although it need not have a zero DC value.

For the linear case, there must be one controlling voltage source and its name is followed by the gain. For the nonlinear case see Chapter 4 for describing the controlling polynomial.

## G Voltage-Controlled Current Source

General forms

G<*name*> <+ *node*> <− *node*>
\+ <+ *controlling node*> <−*controlling node*>
\+ <*transconductance*>
G<*name*> <+ *node*> <− *node*> POLY(<*value*>)
\+ < <+ *controlling node*> <− *controlling node* >*
\+ < <*polynomial coefficient value*> >*

*Examples*

```
GBUFF     1   2 10 11 1.0
GAMP     13 0POLY(1) 26 0 500
GNONLIN 100 101  POLY(2) 3 0 4 0 0.0 13.6 0.2 0.005
```

The first form and the first two examples apply to the linear case. The second form and the last example are for the nonlinear case. POLY(<*value*>) specifies the number of dimensions of the polynomial. The number of pairs of controlling nodes must be equal to the number of dimensions.

The (+) and (−) nodes are the output nodes. A positive current flows from the (+) node through the source to the (−) node. The (+ controlling) and (− controlling) nodes are in pairs and define a set of voltages. A particular node may appear more than once, and the output and controlling nodes need not be different.

For the linear case, there are two controlling nodes and these are followed by the transconductance. For the nonlinear case see Chapter 4 for describing the controlling polynomial.

## H Current-Controlled Voltage Source

General forms

H<*name*> <+ *node*> <−*node*>
\+ <*controlling V device name*> <*transresistance*>
H<*name*> <+*node*> <−*node*> POLY(<*value*>)
\+ <*controlling V device name*>*
\+ < <*polynomial coefficient value*> >*

*Examples*

```
HSENSE    1   2 VSENSE 10.0
HAMP     13   0 POLY(1) VIN 500
HNONLIN 100 101  POLY(2) VCNTRL1 VCINTRL2 0.0 13.6 0.2 0.005
```

The first form and the first two examples apply to the linear case. The second form and the last example are for the nonlinear case. POLY(<*value*>) specifies the number of dimensions of the polynomial. The number of controlling voltage sources must be equal to the number of dimensions.

The (+) and (−) nodes are the output nodes. Positive current flows from the (+) node through the source to the (−) node. The current through the controlling voltage source determines the output voltage. The controlling source must be an independent voltage source (V device), though it need not have a zero DC value.

For the linear case, there must be one controlling voltage source and its name is followed by the transresistance. For the nonlinear case see Chapter 4 for describing the controlling polynomial.

## I Independent Current Source

General forms

I<*name*> <+ *node*> <−*node*>
+ [[DC] <*value*>]
+ [AC <*magnitude value*> [*phase value*]]
+ [*transient specification*]

*Examples*

```
IBIAS   13  0 2.3mA
IAC      2  3 AC .001
IACPHS  2  3 AC .001 90
IPULSE  1  0 PULSE(-1mA 1mA 2nS 2nS 2nS 50nS 100nS)
I3      26  77 DC .002 AC 1 SIN(.002 .002 1.5MEG)
```

This element is a current source. Positive current flows from the (+) node through the source to the (−) node. The default value is zero for the DC, AC, and transient values. None, any, or all of DC, AC, and transient values may be specified. The AC phase value is in degrees.

If present, [*transient specification*] must be one of:

EXP <*parameters*>
PULSE <*parameters*>
PWL <*parameters*>
SFFM <*parameters*>
SIN <*parameters*>

These are individually described in Chapter 9.

## J Junction FET

General forms

J<*name*> <*drain node*> <*gate node*> <*source node*>
+ <*model name*> [*area value*]

*Examples*

```
JIN 100 1 0 JFAST
J13 22 14 23 JNOM 2.0
```

| Model Parameters (see .MODEL statement) | | Default value | Units |
|---|---|---|---|
| VTO | threshold voltage | −2.0 | volt |
| BETA | transconductance coefficient | 1E−4 | $amp/volt^2$ |
| LAMBDA | channel-length modulation | 0 | $volt^{-1}$ |
| RD | drain ohmic resistance | 0 | ohm |
| RS | source ohmic resistance | 0 | ohm |
| IS | gate p-n saturation current | 1E−14 | amp |
| PB | gate p-n potential | 1 | volt |
| CGD | gate-drain zero-bias p-n capacitance | 0 | farad |
| CGS | gate-source zero-bias p-n capacitance | 0 | farad |
| FC | forward-bias depletion capacitance coefficient | .5 | |
| VTOTC | VTO temperature coefficient | 0 | volt/°C |
| BETATCE | BETA exponential temperature coefficient | 0 | %/°C |
| KF | flicker noise coefficient | 0 | |
| AF | flicker noise exponent | 1 | |

The JFET is modeled as an intrinsic FET with an ohmic resistance (RD/area) in series with the drain, and with another ohmic resistance (RS/area) in series with the source, and with another ohmic resistance (RG) in series with the gate. Positive current is current flowing into a terminal. [*area value*] is the relative device area and defaults to 1.

## K **Inductor Coupling (transformer core)**

General forms

K<*name*> L<*inductor name*> <L<*inductor name*> >*
\+ <*coupling value*>
K<*name*> <L<*inductor name*> >* <*coupling value*>
\+ <*model name*> [*size value*]

*Examples*

```
KTUNED L3OUT L4IN .8
KTRNSFRM LPRIMARY LSECNDRY .99
KXFRM L1 L2 L3 L4 .98 KPOT_3C8
```

| Model Parameters (nonlinear magnetics only) | | Default value | Units |
|---|---|---|---|
| AREA | mean magnetic cross-section | .1 | $cm^2$ |
| PATH | mean magnetic path length | 1 | cm |
| GAP | effective air-gap length | 0 | cm |
| PACK | pack (stacking) factor | 1 | |
| MS | magnetization saturation | 1E + 6 | amp/meter |
| ALPHA | mean field parameter | .001 | |
| A | shape parameter | 1000 | amp/meter |
| C | domain wall flexing constant | .2 | |
| K | domain wall pinning constant | 500 | |

K*<name>* couples two or more inductors. Using the "dot" convention, place a "dot" on the first node of each inductor. In other words, given:

```
I1 1 0 AC 1mA
L1 1 0 10uH
L2 2 0 10uH
R2 2 0 .1
K12 L1 L2 .9999
```

the current through L2 will be in the opposite direction as the current through L1. The polarity is determined by the order of the nodes in the L device(s) and not by the order of inductors in the K statement.

*<coupling value>* is the "coefficient of mutual coupling" which must be between 0 and 1. Note that iron-core transformers have a very high coefficient of coupling, greater than .999 in many cases.

If *<model name>* is present, four things change:

The mutual coupling inductor becomes a nonlinear, magnetic core device.

The magnetic core's B-H characteristics are analyzed using the Jiles-Atherton model.

The inductors become "windings," so the number specifying inductance now specifies the "number of turns."

The list of coupled inductors may be just one inductor.

A model statement is required to specify the model parameters.

[*size value*] defaults to 1 and scales the magnetic cross-section. It is intended to represent the number of lamination layers, so only one model statement is needed to each lamination type.

PSpice uses the Jiles-Atherton model to analyze the B-H curve of the magnetic core, and calculate values for inductance and flux for each of the "windings."

## L Inductor

General forms

L<*name*> <*+node*> <*−node*> [*model name*] <*value*>
+ [IC = <*initial value*>]

*Examples*

```
LLOAD  15   0 20mH
L2      1   2 .2E-6
LCHOKE  3  42 LMOD .03
LSENSE  5  12 2UH IC=2mA
```

| Model Parameters (see .MODEL statement) | | Default value | Units |
|---|---|---|---|
| L | inductance multiplier | 1 | |
| IL1 | linear current coefficient | 0 | $amp^{-1}$ |
| IL2 | quadratic current coefficient | 0 | $amp^{-2}$ |
| TC1 | linear temperature coefficient | 0 | $°C^{-1}$ |
| TC2 | quadratic temperature coefficient | 0 | $°C^{-2}$ |

The (+) and (−) nodes define the polarity meant when the inductor has a positive voltage across it. Also, positive current flows from the (+) node through the inductor to the (−) node.

If [*model name*] is left out, then <*value*> is the inductance in henries.

If [*model name*] is specified, then the inductance is given by the formula

$$\langle value \rangle \cdot L \cdot (1 + IL1 \cdot I + IL2 \cdot I^2) \cdot (1 + TC1 \cdot (T - Tnom) + TC2 \cdot (T - Tnom)^2)$$

<*value*> is normally positive (though it can be negative, but **not** zero). "Tnom" is the nominal temperature (set with TNOM option).
<*initial value*> is the initial guess for the current through the inductor during the bias-point calculation.

## M MOSFET

General forms

M<*name*> <*drain node*> <*gate node*> <*source node*>
+ <*bulk/substrate node*> <*model name*>
+ [L = <*value*>] [W = <*value*>]
+ [AD = <*value*>] [AS = <*value*>]
+ [PD = <*value*>] [PS = <*value*>]

\+ [NRD = <*value*>] [NRS = <*value*>]
\+ [NRG = <*value*>] [NRB = <*value*>]

*Examples*

```
M1 14 2  13    0 PNOM L=25u W=12u
M13 15 3   0    0 PSTRONG
M2A  0 2 100 100 NWEAK L=33u W=12u
+AD=288p AS=288p PD=60u PS=60u NRD=14 NRS=24 NRG=10
```

| Model Parameters (see .MODEL statement) | | Default value | Units |
|---|---|---|---|
| LEVEL | model type (1, 2, or 3) | 1 | |
| L | channel length | DEFL | meter |
| W | channel width | DEFW | meter |
| LD | lateral diffusion (length) | 0 | meter |
| WD | lateral diffusion (width) | 0 | meter |
| VTO | zero-bias threshold voltage | 0 | volt |
| KP | transconductance | 2E − 5 | amp/volt$^2$ |
| GAMMA | bulk threshold parameter | 0 | volt$^{1/2}$ |
| PHI | surface potential | .6 | volt |
| LAMBDA | channel-length modulation (LEVEL = 1 or 2) | 0 | volt$^{-1}$ |
| RD | drain ohmic resistance | 0 | ohm |
| RS | source ohmic resistance | 0 | ohm |
| RG | gate ohmic resistance | 0 | ohm |
| RB | bulk ohmic resistance | 0 | ohm |
| RDS | drain-source shunt resistance | infinite | ohm |
| RSH | drain, source diffusion sheet resistance | 0 | ohm/square |
| IS | bulk p-n saturation current | 1E − 14 | amp |
| JS | bulk p-n saturation current/area | 0 | amp/meter$^2$ |
| PB | bulk p-n potential | .8 | volt |
| CBD | bulk-drain zero-bias p-n capacitance | 0 | farad |
| CBS | bulk-source zero-bias p-n capacitance | 0 | farad |
| CJ | bulk p-n zero-bias bottom capacitance/area | 0 | farad/meter$^2$ |
| CJSW | bulk p-n zero-bias perimeter capacitance/length | 0 | farad/meter |
| MJ | bulk p-n bottom grading coefficient | .5 | |
| MJSW | bulk p-n sidewall grading coefficient | .33 | |
| FC | bulk p-n forward-bias capacitance coefficient | .5 | |
| CGSO | gate-source overlap capacitance/channel width | 0 | farad/meter |
| CGDO | gate-drain overlap capacitance/channel width | 0 | farad/meter |
| CGBO | gate-bulk overlap capacitance/channel length | 0 | farad/meter |
| NSUB | substrate doping density | 0 | 1/cm$^3$ |
| NSS | surface state density | 0 | 1/cm$^2$ |
| NFS | fast surface state density | 0 | 1/cm$^2$ |
| TOX | oxide thickness | infinite | meter |
| TPG | gate material type:<br>+1 = opposite of substrate<br>−1 = same as substrate<br>0 = aluminum | +1 | |
| XJ | metallurgical junction depth | 0 | meter |

| Model Parameters (see .MODEL statement) | | Default value | Units |
|---|---|---|---|
| UO | surface mobility | 600 | $cm^2$/volt·sec |
| UCRIT | mobility degradation critical field (LEVEL = 2) | 1E4 | volt/cm |
| UEXP | mobility degradation exponent (LEVEL = 2) | 0 | |
| UTRA | (not used) mobility degradation transverse field coefficient | | |
| VMAX | maximum drift velocity | 0 | meter/sec |
| NEFF | channel charge coefficient (LEVEL = 2) | 1 | |
| XQC | fraction of channel charge attributed to drain | 1 | |
| DELTA | width effect on threshold | 0 | |
| THETA | mobility modulation (LEVEL = 3) | 0 | $volt^{-1}$ |
| ETA | static feedback (LEVEL = 3) | 0 | |
| KAPPA | saturation field factor (LEVEL = 3) | .2 | |
| KF | flicker noise coefficient | 0 | |
| AF | flicker noise exponent | 1 | |

The MOSFET is modeled as an intrinsic MOSFET with ohmic resistances in series with the drain, source, gate, and bulk (substrate). There is also a shunt resistance (RDS) in parallel with the drain-source channel. Positive current is current flowing into a terminal (for example, positive drain current flows from the drain through the channel to the source).

L and W are the channel length and width. L is decreased by twice LD to get the effective channel length. W is decreased by twice WD to get the effective channel width. L and W can be specified on the device, the model, or on the .OPTION statement. The value on the device supersedes the value on the model which supersedes the value on the .OPTION statement.

AD and AS are the drain and source diffusion areas. PD and PS are the drain and source diffusion perimeters. The drain-bulk and source-bulk saturation currents can be specified either by JS, which is multiplied by AD and AS, or by IS, which is an absolute value. The zero-bias depletion capacitances can be specified by CJ, which is multiplied by AD and AS, and by CJSW, which is multiplied by PD and PS. Or they can be set by CBD and CBS, which are absolute values.

NRD, NRS, NRG, and NRB are the relative resistivities of the drain, source, gate, and substrate in squares. These parasitic (ohmic) resistances can be specified either by RSH, which is multiplied by NRD, NRS, NRG, and NRB respectively, or by RD, RS, RG, and RB, which are absolute values.

PD and PS default to 0. NRD and NRS default to 1. NRG and NRB default to 0. Defaults for L, W, AD, and AS may be set in the .OPTIONS statement. If AD or AS defaults are not set, they also default to 0. If L or W defaults are not set, they default to 100u.

# Q Bipolar Transistor

General forms

Q<*name*> <*collector node*> <*base node*> <*emitter node*>
+ [*substrate node*] <*model name*> [*area value*]

*Examples*

```
Q1  14 2 13 PNPNOM
Q13 15 3  0 1 NPNSTRONG 1.5
```

| Model Parameters (see .MODEL statement) | | Default value | Units |
|---|---|---|---|
| IS | p-n saturation current | 1E − 16 | amp |
| BF | ideal maximum forward beta | 100 | |
| NF | forward current emission coefficient | 1 | |
| VAF (VA) | forward Early voltage | infinite | volt |
| IKF (IK) | corner for forward beta high-current roll-off | infinite | amp |
| ISE (C2) | base-emitter leakage saturation current | 0 | amp |
| NE | base-emitter leakage emission coefficient | 1.5 | |
| BR | ideal maximum reverse beta | 1 | |
| NR | reverse current emission coefficient | 1 | |
| VAR (VB) | reverse Early voltage | infinite | volt |
| IKR | corner for reverse beta high-current roll-off | infinite | amp |
| ISC (C4) | base-collector leakage saturation current | 0 | amp |
| NC | base-collector leakage emission coefficient | 2.0 | |
| RB | zero-bias (maximum) base resistance | 0 | ohm |
| RBM | minimum base resistance | RB | ohm |
| IRB | current at which Rb falls halfway to RBM | infinite | amp |
| RE | emitter ohmic resistance | 0 | ohm |
| RC | collector ohmic resistance | 0 | ohm |
| CJE | base-emitter zero-bias p-n capacitance | 0 | farad |
| VJE (PE) | base-emitter built-in potential | .75 | volt |
| MJE (ME) | base-emitter p-n grading factor | .33 | |
| CJC | base-collector zero-bias p-n capacitance | 0 | farad |
| VJC (PC) | base-collector built-in potential | .75 | volt |
| MJC (MC) | base-collector p-n grading factor | .33 | |
| XCJC | fraction of Cbc connected internal to Rb | 1 | |
| CJS (CCS) | collector-substrate zero-bias p-n capacitance | 0 | farad |
| VJS (PS) | collector-substrate built-in potential | .75 | volt |
| MJS (MS) | collector-substrate p-n grading factor | 0 | |
| FC | forward-bias depletion capacitor coefficient | .5 | |
| TF | ideal forward transit time | 0 | sec |
| XTF | transit time bias dependence coefficient | 0 | |
| VTF | transit time dependency on Vbc | infinite | volt |
| ITF | transit time dependency on Ic | 0 | amp |
| PTF | excess phase @ $1/(2\pi \cdot TF)$Hz | 0 | degree |

| Model Parameters (see .MODEL statement) | | Default value | Units |
|---|---|---|---|
| TR | ideal reverse transit time | 0 | sec |
| EG | bandgap voltage (barrier height) | 1.11 | eV |
| XTB | forward and reverse beta temperature coefficient | 0 | |
| XTI (PT) | IS temperature effect exponent | 3 | |
| KF | flicker noise coefficient | 0 | |
| AF | flicker noise exponent | 1 | |

The bipolar transistor is modeled as an intrinsic transistor with ohmic resistances in series with the collector (RC/area), the base (value varies with current) and with the emitter (RE/area). The substrate node is optional, and if not specified it defaults to ground. Positive current is current flowing into a terminal. [*area value*] is the relative device area and defaults to 1. For those model parameters which have alternate names, such as VAF and VA (the alternate name is indicated with parentheses), either name may be used.

The parameters ISE (C2) and ISC (C4) may be set to be greater than 1. In this case, they are interpreted as multipliers of IS instead of absolute currents; that is, if ISE > 1, then it is replaced by ISE·IS. The same applies for ISC.

## R Resistor

General forms

R<*name*> <+*node*> <− *node*> [*model name*] <*value*>

*Examples*

```
RLOAD 15 0 2K
R2    1 2 2.4E4
```

| Model Parameters (see .MODEL statement) | | Default value | Units |
|---|---|---|---|
| R | resistance multiplier | 1 | |
| TC1 | linear temperature coefficient | 0 | $°C^{-1}$ |
| TC2 | quadratic temperature coefficient | 0 | $°C^{-2}$ |
| TCE | exponential temperature coefficient | 0 | %/°C |

The (+) and (−) nodes define the polarity meant when the resistor has a positive voltage across it. Positive current flows from the (+) node through the resistor to the (−) node.

If [*model name*] is included and TCE (in the model) **is not specified,** then the resistance is given by the formula

$$<value>\cdot R\cdot(1 + TC1\cdot(T - Tnom) + TC2\cdot(T - Tnom)^2)$$

If [*model name*] is included and TCE (in the model) **is specified,** then the resistance is given by the formula

$$<value> \cdot R \cdot 1.01^{TCE \cdot (T - Tnom)}$$

<*value*> is normally positive (though it can be negative, but **not** zero). "Tnom" is the nominal temperature (set with TNOM option).

*Noise:* Noise is calculated assuming a 1 hertz bandwidth. The resistor generates thermal noise with the following spectral power density (per unit bandwidth):

$$i^2 = 4 \cdot k \cdot T/resistance$$

## S Voltage-Controlled Switch

General forms

S<*name*> <+ *switch node*> <−*switch node*>
\+ <+ *controlling node*> <−*controlling node*>
\+ <*model name*>

*Examples*

```
S12     13 17  2 0 SMOD
SRESET  5   0 15 3 RELAY
```

| Model Parameters (see .MODEL statement) | | Default value | Units |
|---|---|---|---|
| RON | "on" resistance | 1 | Ohm |
| ROFF | "off" resistance | 1E + 6 | Ohm |
| VON | control voltage for "on" state | 1 | Volt |
| VOFF | control voltage for "off" state | 0 | Volt |

The voltage-controlled switch is a special kind of voltage-controlled resistor. The resistance between <+ *switch node*> and <− *switch node*> depends on the voltage between <+ *controlling node*> and <− *controlling node*>. The resistance varies continuously between RON and ROFF.

RON and ROFF must be greater than zero and less than 1/GMIN.

A resistance of 1/GMIN is connected between the controlling nodes to keep them from floating. See the .OPTION card to change GMIN.

We have chosen this model for a switch to try to minimize numerical problems. However, there are a few things to keep in mind:

- With double precision numbers PSpice can handle only a dynamic range of about 12 decades. So, we do not recommend making the ratio of ROFF to RON greater than 1E + 12.
- Similarly, we do not recommend making the transition region too narrow. Remember that in the transition region the switch has gain. The narrower the region, the higher the gain and the greater the potential for numerical problems.
- Although very little computer time is required to evaluate switches, during transient analysis PSpice must step through the transition region with a fine enough step size to get an accurate waveform. So, for many transitions you may have long run times from evaluating the other devices in the circuit many times.

## T Transmission Line

General forms

T<*name*> <+ *A port node*> <− *A port node*>
\+ <+ *B port node*> <− *B port node*>
\+ Z0 = <*value*> [TD = <*value*>] [F = <*value*> [NL = <*value*>]]

*Examples*

```
T1 1 2 3 4 Z0=220 TD=115nS
T2 1 2 3 4 Z0=220 F=2.25MEGHz
T3 1 2 3 4 Z0=220 F=4.5MEGHz NL=0.5
```

The transmission line device is a bidirectional, ideal delay line. It has two ports, A and B. The (+) and (−) nodes define the polarity of a positive voltage at a port.

Z0 is the characteristic impedance. The transmission line's length can be specified either by TD, a delay in seconds, or by F and NL, a frequency and a relative wavelength at F. NL defaults to 0.25 (F is then the quarter-wave frequency). Although TD and F are both shown as optional, one of the two must be specified. Examples T1, T2, and T3 all specify the same transmission line.

## V Independent Voltage Source

General forms

V<*name*> <+ *node*> <− *node*>
\+ [[DC] <*value*>]

\+ [AC <*magnitude value*> [*phase value*]]
\+ [*transient specification*]

*Examples*

```
VBIAS  13  0 2.3mV
VAC     2  3 AC .001
VACPHS  2  3 AC .001 90
VPULSE  1  0 PULSE(-1mV 1mV 2nS 2nS 2nS 50nS 100nS)
V3     26 77 DC .002 AC 1 SIN(.002 .002 1.5MEG)
```

This element is a voltage source. Positive current flows from the (+) node through the source to the (−) node. The default value is zero for the DC, AC, and transient values. None, any, or all of DC, AC, and transient values may be specified. The AC phase value is in degrees.

If present, [*transient specification*] must be one of:

EXP <*parameters*>
PULSE <*parameters*>
PWL <*parameters*>
SFFM <*parameters*>
SIN <*parameters*>

These are individually described in Chapter 9.

## W Current-Controlled Switch

General forms

W<*name*> <+ *switch node*> <− *switch node*>
\+ <*controlling V device name*> <*model name*>

*Examples*

```
W12     13 17 VC WMOD
WRESET  5   0 VRESET RELAY
```

| Model Parameters (see .MODEL statement) | | Default value | Units |
|---|---|---|---|
| RON | "on" resistance | 1 | ohm |
| ROFF | "off" resistance | 1E + 6 | ohm |
| ION | control current for "on" state | .001 | amp |
| IOFF | control current for "off" state | 0 | amp |

The current-controlled switch is a special kind of voltage-controlled resistor. The resistance between <+ *switch node*> and <− *switch node*> depends on the current through <*controlling V device name*>. The resistance varies continuously between RON and ROFF.

RON and ROFF must be greater than zero and less than 1/GMIN.

A resistance of 1/GMIN is connected between the controlling nodes to keep them from floating. See the .OPTION card to change GMIN.

We have chosen this model for a switch to try to minimize numerical problems. However, there are a few things to keep in mind:

- With double precision numbers PSpice can handle only a dynamic range of about 12 decades. So, we do not recommend making the ratio of ROFF to RON greater than 1E + 12.
- Similarly, we do not recommend making the transition region too narrow. Remember that in the transition region the switch has gain. The narrower the region, the higher the gain and the greater the potential for numerical problems.
- Although very little computer time is required to evaluate switches, during transient analysis PSpice must step through the transition region with a fine enough step size to get an accurate waveform. So, for many transitions you may have long run times from evaluating the other devices in the circuit many times.

### X Subcircuit Call

General forms

X<*name*> [*node*]* <*subcircuit name*>

*Examples*

```
X12   100 101 200 201 DIFFAMP
XBUFF  13   15 UNITAMP
```

<*subcircuit name*> is the name of the subcircuit's definition (see .SUBCKT statement). There must be the same number of nodes in the call as in the subcircuit's definition. This statement causes the referenced subcircuit to be inserted into the circuit with the given nodes replacing the argument nodes in the definition. It allows you to define a block of circuitry once and then use that block in several places.

Subcircuit calls may be nested. That is, you may have a call to subcircuit A, whose definition contains a call to subcircuit B. The nesting may be to any level, but **must not be circular.** For example, if subcircuit A's definition contains a call to subcircuit B, then subcircuit B's definition must not contain a call to subcircuit A.

# APPENDIX C

# How PSpice Works

The details of the concepts and algorithms of circuit simulation, especially as they relate to SPICE and PSpice, are contained in the thesis:

NAGEL, LAURENCE. *SPICE2: A Computer Program to Simulate Semiconductor Circuits,* Memorandum No. M520 (May 1975).

This is available by sending a check for $30, payable to *The Regents of the University of California,* to this address:

Cindy Manly
EECS/ERL Industrial Liaison Program
497 Cory Hall
University of California
Berkeley, California 94720

This thesis reviews and develops many of the methods that could be used for numerically simulating electronic circuits, and covers the advantages (and pitfalls) of these in great detail.

A much shorter review of how SPICE and PSpice works, using the algorithms from the aforementioned thesis, is

BLUME, WOLFRAM. *Computer Circuit Simulation*, *BYTE*, vol 11 no. 7 (July 1986), page 165.

You should read this article as your introduction to the ''innards'' of SPICE. This will probably satisfy most users as to the details of the algorithms in the program. If not, then read the thesis by Laurence Nagel. Also interesting is a recently published book reviewing these algorithms and their use in a variety of simulators:

WILLIAM J. MCCALLA, "Fundamentals of Computer-Aided Circuit Simulation," Kluwer Academic, 1988.

Note: in the process of publication of the article by Wolfram Blume, an error occurred in Figure 2 of that article. The conductance matrix (as printed) connects R2 between nodes 1 and 0, instead of nodes 2 and 0. The matrix should have been written:

$$\begin{matrix} 1/R2 & 0 & -1/R2 \\ 0 & 1/R1 & -1/R1 \\ -1/R2 & -1/R1 & 1/R1 + 1/R2 \end{matrix}$$

Choosing V0 (node 0) as ground, and setting it to 0 volts, gives three equations:

$$-I = -V2/R2$$

$$+I = V1/R1 - V2/R1$$

$$0 = -V1/R1 + V2 \cdot (1/R1 + 1/R2)$$

These equations reduce to:

$$I = V1/(R1 + R2)$$

$$V2 = V1 \cdot R2/(R1 + R2)$$

and are the results you would expect by examination of the circuit.

# APPENDIX D

# Voltage-Controlled Components

One of the most commonly requested component additions to PSpice is a voltage controlled-resistor or voltage-controlled capacitor. Actually, a small subcircuit definition will provide these functions, as well as a voltage-controlled inductor. These subcircuits should also work with U. C. Berkeley SPICE and other commercial offerings.

The "YX" subcircuit creates a floating, voltage-controlled admittance (remember, admittance is inverse of impedance). It does this by mirroring the voltage at the terminals of the pseudo-component, multiplied by the control voltage to the reference component, which is either a capacitor or a conductance (1/resistance). As a result, the current that flows through the reference component will be proportional to the pseudo-component voltage, as well as to the control voltage. The resulting current is then mirrored to the output terminals.

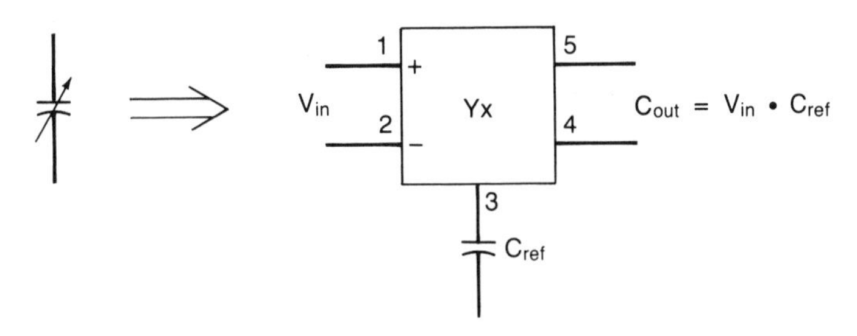


Schematic of "YX" circuit.

```
*Copyright 1987, MicroSim Corporation
*
*Variable admittance: Yout = Yin * V
*            control input: voltage
*             |  capacitor/conductance (connect other lead to ground)
*            / \ |   output: floating admittance (pseudo-component)
*            + - | / \
.subckt yx 1 2 3 4 5
  ecopy 3 6 poly(2) (1,2) (4,5) 0 0 0 0 1
  fout  4 5 vsense 1
  rin   1 2 1G
  vsense 0 6 0
.ends
```

The "ZX" subcircuit uses similar techniques to create a floating, voltage-controlled impedance. In this case the current flowing through the pseudo-component is mirrored directly to the reference component, which is either an inductor or a resistor. The resulting voltage is then mirrored to the output terminals, multiplied by the controlling voltage. As a result, the voltage at the output terminals is proportional to the voltage across the reference component, as well as the control voltage.

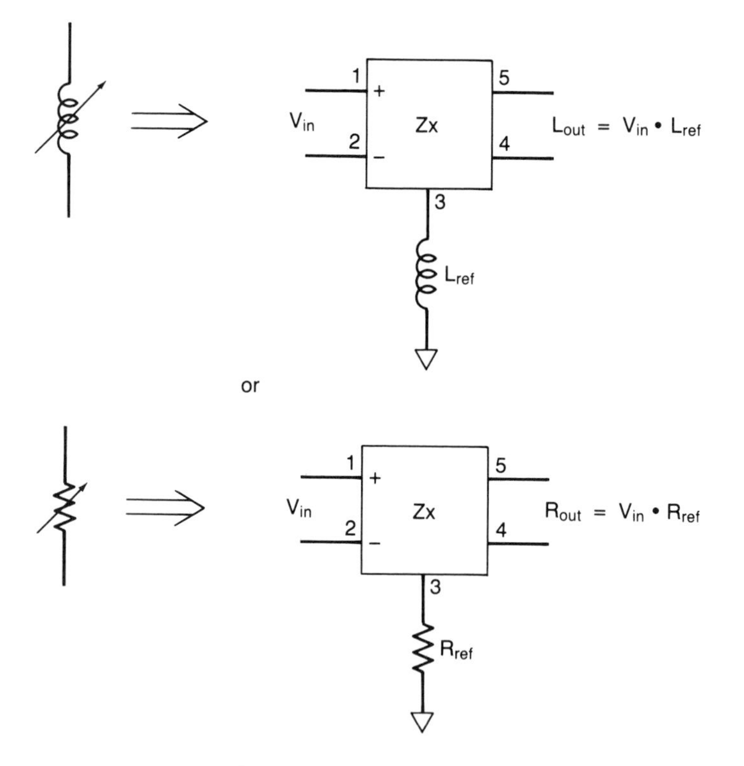


Schematic of "ZX" circuit.

```
*Copyright 1987, MicroSim Corporation
*
*Variable impedance: Zout = Zin * V
*             control input: voltage
*             | inductor/resistor (connect other lead to ground)
*           / \ |  output: floating impedance (pseudo-component)
*           + - | / \
.subckt zx 1 2 3 4 5
  eout   4 6 poly(2) (1,2) (3,0) 0 0 0 0 1
  fcopy  0 3 vsense 1
  rin    1 2 1G
  vsense 6 5 0
.ends
```

# Index

## A



## B



## C

## D

# I



# J



# K



# L



# M

# N



# O

## P



## Q



## R

# S



# T

## U



## V



## W



## X



## Y



## Z

Tear out this card and fill in all necessary information. Then enclose this card with your check or money order only in an envelope and mail to:

Book Distribution Center
PRENTICE HALL
Route 59 at
Brook Hill Drive
West Nyack, New York
10995

**SPICE: A GUIDE TO CIRCUIT SIMULATION AND ANALYSIS USING PSPICE®—Paul W. Tuinenga**

Please send the item(s) checked below. PAYMENT ENCLOSED (Check or money order only.) The Publisher will pay all shipping and handling charges.

____ PSpice® Student Version Disks (2) IBM® PC compatible. $7.00 each set. (83463-0)

____ PSpice® Student Version Disk (1) MAC II® compatible. $6.00 each disk. (83462-2)

____ PSpice® Student Version Disk (1) IBM PS/2 compatible. $6.00 each disk. (83464-8)

NAME ____________________

DEPT. ____________________

SCHOOL ____________________

CITY ____________ STATE ________ ZIP ________

NOTE: PROFESSIONAL/REFERENCE BOOKS ARE TAX DEDUCTIBLE.
Prices subject to change without notice. Please add sales tax for your area.

Dept. 1 D-CFJB-BE(6)


Tear out this card and fill in all necessary information. Then enclose this card with your check or money order only in an envelope and mail to:

Book Distribution Center
PRENTICE HALL
Route 59 at
Brook Hill Drive
West Nyack, New York
10995

**SPICE: A GUIDE TO CIRCUIT SIMULATION AND ANALYSIS USING PSPICE®—Paul W. Tuinenga**

Please send the item(s) checked below. PAYMENT ENCLOSED (Check or money order only.) The Publisher will pay all shipping and handling charges.

____ PSpice® Student Version Disks (2) IBM® PC compatible. $7.00 each set. (83463-0)

____ PSpice® Student Version Disk (1) MAC II® compatible. $6.00 each disk. (83462-2)

____ PSpice® Student Version Disk (1) IBM PS/2 compatible. $6.00 each disk. (83464-8)

NAME ____________________

DEPT. ____________________

SCHOOL ____________________

CITY ____________ STATE ________ ZIP ________

NOTE: PROFESSIONAL/REFERENCE BOOKS ARE TAX DEDUCTIBLE.
Prices subject to change without notice. Please add sales tax for your area.

Dept. 1 D-CFJB-BE(6)